Interspecies Interact

Interspecies Interactions surveys the rapidly developing field of human–animal relations from the late medieval and early modern eras through to the mid-Victorian period. By viewing animals as authentic and autonomous historical agents who had a real impact on the world around them, this book concentrates on an under-examined but crucial aspect of the human–animal relationship: interaction.

Each chapter provides scholarly debate on the methods and challenges of the study of interspecies interactions, and together they offer an insight into the parts that humans and animals have played in shaping each other's lives, as well as encouraging reflection on the directions that human–animal relations may yet take. Beginning with an exploration of Samuel Pepys' often emotional relationships with the many animals that he knew, the chapters cover a wide range of domestic, working, and wild animals and include case studies on carnival animals, cattle, dogs, horses, apes, snakes, sharks, invertebrates. These case studies of human–animal interactions are further brought to life through visual representation, by the inclusion of over 20 images within the book.

From 'sleeve cats' to lion fights, *Interspecies Interactions* encompasses a broad spectrum of relationships between humans and animals. Covering topics such as use, emotion, cognition, empire, status, and performance across several centuries and continents, it is essential reading for all students and scholars of historical animal studies.

Sarah Cockram is Lecturer in History, *c*.1200–1600 at the University of Glasgow. Her publications include *Isabella d'Este and Francesco Gonzaga: Power Sharing at the Italian Renaissance Court* (2013) and a co-edited special issue of the journal *Renaissance Studies* on 'The Animal in Renaissance Italy'.

Andrew Wells is a postdoctoral researcher (Wissenschaftlicher Mitarbeiter) in the Graduate School of the Humanities (GSGG) at the Georg-August-Universität Göttingen, Germany. He has published essays in *History Compass*, the *Journal of British Studies*, and *History of European Ideas*. His forthcoming book explores the interactions of racial and sexual concepts and identities in eighteenth-century British culture.

Interspecies Interactions

Animals and Humans between the
Middle Ages and Modernity

**Edited by
Sarah Cockram and
Andrew Wells**

Routledge
Taylor & Francis Group

LONDON AND NEW YORK

First published 2018
by Routledge
2 Park Square, Milton Park, Abingdon, Oxon OX14 4RN

and by Routledge
711 Third Avenue, New York, NY 10017

Routledge is an imprint of the Taylor & Francis Group, an informa business

British Library Cataloguing-in-Publication Data
A catalogue record for this book is available from the British Library

Library of Congress Cataloging-in-Publication Data
Names: Cockram, Sarah D. P., editor. | Wells, Andrew, 1979– editor.
Title: Interspecies interactions: animals and humans between the middle ages and modernity / [edited by] Sarah Cockram and Andrew Wells.
Description: Milton Park, Abingdon, Oxon; New York, NY: Routledge, 2017. | Includes bibliographical references and index.
Identifiers: LCCN 2017011631| ISBN 9781138189713 (hbk: alk. paper) | ISBN 9781138189720 (pbk: alk. paper) | ISBN 9781315109299 (ebk)
Subjects: LCSH: Human–animal relationships—Europe—History.
Classification: LCC QL85 .I58 2017 | DDC 591.5—dc23
LC record available at https://lccn.loc.gov/2017011631

ISBN: 978-1-138-18971-3 (hbk)
ISBN: 978-1-138-18972-0 (pbk)
ISBN: 978-1-315-10929-9 (ebk)

Typeset in Bembo
by codeMantra

For Katrin
To Tara

Contents

Figures

Contributors

László Bartosiewicz has been researching animal bones from archaeological sites for thirty-five years. His main areas of interest include animal-human relationships in the past and the past use of animals as media in human communication. While specialised in the study of osteological remains he considers multidisciplinarity indispensable in understanding the cultural roles played by animals. He has taught archaeozoology at the Universities of Budapest (Hungary) and Edinburgh (UK) and is currently working at Stockholm University (Sweden).

Ido Ben-Ami is a PhD Candidate in the Zvi Yavetz School of Historical Studies at Tel Aviv University. His main research interests include the Ottoman Empire in the early modern period and the history of emotions. His doctoral dissertation focuses on emotions and human-animal relations in early modern Istanbul.

Sarah Cockram is Lecturer in History, c.1200–1600 at the University of Glasgow. Prior to this she held a Leverhulme Early Career Fellowship at the University of Edinburgh for her project 'Courtly Creatures: Animals and Image Construction at the Italian Renaissance Court'. Her research examines society, culture and power in medieval and early modern Europe, with a focus on animal studies, gender, and the Renaissance. Sarah's publications include *Isabella d'Este and Francesco Gonzaga: Power Sharing at the Italian Renaissance Court* (2013) and a special issue on animals of the journal *Renaissance Studies* (co-edited with Stephen Bowd), to which Sarah has contributed an article on the interspecies understanding between Renaissance animal handlers and their exotic charges such as giraffes, lions, elephants, rhinos and cheetahs.

Helen Cowie is Lecturer in History at the University of York. Her research focuses on the history of animals and the history of natural history. She is author of *Conquering Nature in Spain and its Empire, 1750–1850* (2011), *Exhibiting Animals in Nineteenth-Century Britain: Empathy, Education, Entertainment* (2014) and *Llama* (2017).

Pia Cuneo is Professor of Art History at the University of Arizona. Her work focuses on the nexus between visual culture and hippology in early modern Germany. She competes locally in dressage and both her horses are rescue animals.

Peter Edwards is Professor Emeritus of Early Modern British Social and Economic History at the University of Roehampton and has written extensively on the multi-functional role of horses in pre-modern society. His publications include *The Horse Trade of Tudor and Stuart England* (1988/ reprinted 2004); *Horse and Man in Early Modern England* (2007); and with Prof. Elspeth Graham (eds), *The Horse as Cultural Icon: The Real and the Symbolic Horse in the Early Modern World* (2011). He is currently writing a book on William Cavendish, First Earl of Devonshire and his horses.

Erica Fudge is Professor of English Studies at the University of Strathclyde and Director of the British Animal Studies Network. Recent articles on the early modern period have looked at animal faces, the concept of 'choreography' and its value for thinking about past human-animal interactions, and the animal-made object, and she is currently completing a study of human-livestock relations in the early seventeenth century. Erica has also written widely about the implications of the focus on animals for the discipline of history, with recent essays in the journal *History & Theory*, and in the *Oxford Handbook on Animal Studies*. In addition to this scholarly work, she has written for a popular audience with the books *Animal* (2004) and *Pets* (2008) and has had essays published in *History Today* magazine.

Dominik Hünniger is Managing Director of the Lichtenberg-Kolleg – the Göttingen Institute for Advanced Study. He is a cultural historian with a special interest in eighteenth-century environmental, medical and natural history as well as the history of universities and scholarship. His current research project focuses on the interconnections among cameralism, natural history and (colonial) exploration at the end of the eighteenth century using the example of the Danish entomologist Johann Christian Fabricius (1745–1808). This project also features research on the development of entomology as a discipline vis-à-vis the material culture of zoological collections and natural history collecting practices.

Krista Maglen is an Associate Professor in the History Department at Indiana University, Bloomington. Her current book project seeks to examine historic encounters and interactions with dangerous animals in Australian history. Her first book, *The English System: Quarantine, Immigration, and the Making of a Port Sanitary Zone* (2014) was shortlisted for the Royal Historical Society's Whitfield Prize.

Karen Raber is Professor of English at the University of Mississippi. She is the author of *Animal Bodies, Renaissance Culture* (2013), and series editor for Routledge's 'Perspectives on the Non-Human in Literature and Culture'.

She has edited or co-edited numerous volumes and written more than thirty chapters and essays on animals, ecostudies, and gender and women writers. She is currently working on a monograph on the materiality of meat in early modern England and Europe.

Harriet Ritvo is the Arthur J. Conner Professor of History at the Massachusetts Institute of Technology. She is the author of *The Dawn of Green: Manchester, Thirlmere, and Modern Environmentalism* (2009), *The Platypus and the Mermaid, and Other Figments of the Classifying Imagination* (1997), *The Animal Estate: The English and Other Creatures in the Victorian Age* (1987), and *Noble Cows and Hybrid Zebras: Essays on Animals and History* (2010). Her articles and reviews have appeared in a wide range of periodicals, including *The London Review of Books, Science, Daedalus, The American Scholar, Technology Review,* and *The New York Review of Books*, as well as scholarly journals in several fields. Her current research concerns wildness and domestication.

Andrew Wells is a postdoctoral researcher (*Wissenschaftlicher Mitarbeiter*) in the Graduate School of the Humanities (GSGG) at the Georg-August-Universität Göttingen, Germany. He has published essays in *History Compass*, the *Journal of British Studies*, and *History of European Ideas*. His forthcoming book explores the interactions of racial and sexual concepts and identities in eighteenth-century British culture. He is currently working on a study of freedom in cities across the Anglophone Atlantic world.

Acknowledgements

This collection of essays has its origin in an interdisciplinary conference, "Between Apes and Angels: Human and Animal in the Early Modern World", held at the University of Edinburgh between 4 and 6 December 2014. The conference attracted participants from across Europe, North America, and the Middle East, and the editors would like to thank all those who attended or showed interest in the gathering and helped us to publicise the event. It was an especially difficult task to decide which speakers to invite from amongst all the high quality submissions, and the editors would like to express their gratitude to all those who responded to the call for papers. In addition to the contributors to this volume, we especially want to thank those who spoke and presided at the conference, particularly Alison Acton, Emily Aleev-Snow, Richard Almond, Miranda Anderson, Benjamin Arbel, Magdalena Bayreuther, Emily Brady, Simona Cohen, Louise Hill Curth, Karen Edwards, Frances Gage, Andrew Gardiner, Angelica Groom, Tom Johnson, Fiona Salvesen Murrell, Neil Pemberton, Laura Peterson, Pauline Phemister, Alan Ross, Ingrid Tague, Elisabeth Wallmann, and Susan Wiseman. For financially supporting the meeting, the editors would like to express their heartfelt appreciation to the Wellcome Trust, the School of History, Classics and Archaeology at the University of Edinburgh, and the Society for Renaissance Studies. The gathering could not have run so smoothly without the generous assistance of Laura Drummond, Karen Howie, Elaine Philip and our student helpers, Alexis Dorsey, Ben Garlick, Jie Li, Haley Pereira, and Rossana Zetti. The conference was such an intellectually exhilarating and convivial gathering that we did not want the conversations there to end. It is pleasing therefore to have them continue, and move into the future, through this book.

Behind the meeting in December 2014 was a research group, also named "Between Apes and Angels", based in the College of Humanities and Social Sciences at the University of Edinburgh. We would like to express our gratitude to the College and to the Royal Society of Edinburgh for financially supporting the activities of this group. Profound thanks go to our collaborators in the research group, Stephen Bowd and Jill Burke, whose good cheer, friendship, sage advice and inspiration have accompanied us throughout all stages of this work and beyond. Thanks also go to Andrew Kitchener and

Andrew Gardiner, and to Catriona Paul and Andrea Roe, whose hard work exposed the richness of animal-related collections that had been hitherto hidden in storerooms and vaults.

For their support and encouragement, the editors would like to express their gratitude to current and former colleagues at Edinburgh, Glasgow, and Göttingen. Andrew's particular thanks are due to Aleksandra Bovt, Jens Elze, Gösta Gabriel, Julia Hauser, Sabine Heerwart, Nele Hoffman, Jennifer Hübel, Florian Kappeler, Anne-Marie Kilday, Erika Manders, Claudia Nickel, Daniele Panizza, Anna Stuhldreher, Christiana Werner, and all the members of the Nachwuchsgruppe 'Wissen'. Sarah's go to Karen Lury, Erica Fudge, Debra Strickland, Julia Smith, Maggie Reilly, Anne Dulau Beveridge, Claire Daniel, Bob MacLean, Clare Knottenbelt, Stewart Mercer and Lizanne Henderson for thought-provoking conversations about animals and for the opening up of exciting possibilities. Sarah would also like to express special appreciation to her students at the Universities of Glasgow and Edinburgh, for their insightful perspectives on the human-animal relationship past and present. Interaction in the classroom and on visits to galleries and collections is a source of great intellectual reward and pleasure.

We would also like to offer special thanks to all at Routledge, especially Catherine Aitken and Morwenna Scott, and, above all, to Laura Pilsworth. We extend our gratitude too to this book's anonymous readers who offered wisdom and generous suggestions.

We owe a tremendous amount to the support and patience of friends and family. To our parents, siblings and family members of the families Wells, Berndt, Blanche, Cockram, Duncan, and Charles and especially to our children Daniel, Eve, and Jamie, we express our unending gratitude. We also want to thank the friends that have accompanied this project, including Hannah Christmas, Sheila Docherty, Colin Massy, Chelsea Sambells, Mirjam Schaub, Katie Hunter, and the Tomkins, King and McKerracher families.

Our final, most profound and affectionate thanks must go to our partners, Katrin Berndt and Paul Cockram, without whom this book could not have been started, let alone completed.

Foreword

Erica Fudge

A core tradition of western thinking has shown us how apparently easy it is to separate humans from animals; to assume that 'we' live in a distinct realm from 'them'; and that 'they' are different from (i.e. inferior to) us because of what we supposedly possess and they lack. Jacques Derrida, as is now familiar, began a deconstruction of such assumptions in *The Animal That Therefore I Am* when he noted the 'nonfinite number' of ways in which human exceptionalism has been articulated in philosophy. Among those faculties or possessions that have been used to express the special and superior nature of *homo sapiens* he listed 'speech or reason, the *logos*, history, laughing, mourning, burial, the gift, etc.' (with that final, abbreviated, 'etc.' ironically marking the unfinished nature of the already absurd list he had provided).[1] Despite the lack of agreement as to 'the definition of *the* limit presumed to separate man in general from the animal in general', Derrida argued that '*all* philosophers, have judged that limit to be single and indivisible, considering that on the other side of that limit there is an immense group, a single and fundamentally homogeneous set that one has the right, the theoretical or philosophical right, to distinguish and mark as opposite, namely, the set of the Animal in general, the Animal spoken of in the general singular'.[2] His claim that '*all* philosophers' have thought this might be challenged, but Derrida is surely right to signal just how influential this tradition in western thought is. In the light of such a tradition, to think about interspecies interactions as this collection does is to begin to review the claims of human exceptionalism and to embrace the possibility that human and animal lives are not so distinct after all.

In terms of our understanding of the past, one way of addressing this possibility would be to trace the histories of the ways in which humans and animals have lived alongside each other: in the domestic setting of the home; at work on the farm, in the stable, in the laboratory; in the contested spaces of the wild. But such a perspective has its limits. Contemplating living 'alongside' animals might allow for an assessment that emphasises how life is experienced in the presence of animals, but it doesn't necessarily allow that co-existence to be constitutive – to have an active role. The concept of lives lived 'alongside' each other allows for the possibility of seeing them running parallel, but remaining distinct, like lanes on a dual carriageway. Better than

thinking about lives lived 'alongside', perhaps, is the idea of human and animal lives being entwined – meeting, crossing over, folding into each other.
Here the simile of the dual carriageway obviously does not work: if those
lanes met, crossed-over, folded into one another crashes would occur; pile-
ups would be inevitable. As a plan for a road system it would be a total failure.
But when thinking about human relations with animals, the conception of
'entwinement' might allow us to see things that 'alongside' misses, or avoids.
There are crashes to be glimpsed, to be sure. But, focusing on 'entwinement'
might also bring to light other interlacings in which outcomes are more positive. As well as revealing collisions, it can also show intimacy and wonder.

The essays that make up *Interspecies Interactions* offer a variety of ways of
viewing the possible value that paying attention to moments of entwinement
might have to our understanding of the past. The moments examined are all
too often bad for the animals – they are baited, eaten, whipped, sold, hunted,
etc. But they are not always good for the humans either: watching can slide
into being brutalised; eating can also be a question of being consumed.
Recognising that there are snakes out there or that a horse's infertility might
lead to human bankruptcy reveals the frailty of human agency as much as it
might also tell us a great deal about humanity's attempts to control its world.
But control does not always produce domination: to recognise an animal's
sentience is also to begin to contemplate its capacity for suffering, and so observation – so often an objectifying act – can also produce a new assessment
of the subject status of the observed thing. The baiting of a lion is a cruel
spectacle, but it is also, perhaps, the beginning of a new way of thinking.

The fact that the essays here cover a wide geographical range – from the
Ottoman Empire to Australia, from Hungary to the New World and use a
wide range of sources – newsprint, painting, street parades, scientific documents, philosophical debates – shows the widespread significance of animals.
This is not a revelation: it has long been recognised that humans have always
lived alongside and come to rely on other species and have also used them
to think with. What this collection offers that is notable, however, is in its
historical range. As the editors signal in their introduction: no comfortable
narrative of improvement can be traced here. We might see scientific, moral,
and legal changes, but what we also come to recognise – and so are asked
implicitly to address – is that despite developments in comprehension, despite
acknowledgements of sentience across species boundaries, human actions are
often performed with assumptions of superiority, difference, right. A nation
and a species might be intertwined, but that does not mean that that species
is not also a commodity.

Derrida proposed that since the beginnings of western philosophy '*all*
philosophers' have colluded in constructing a singular creature called 'the
Animal' – a creature whose specific characteristics have been eradicated and
whose real potential for suffering has been erased. His aim was to ask philosophy to reassess its own past; to question some of its own assumptions and
move forward from there. Perhaps one of the most important outcomes of

reading the essays that follow here will be that they ask readers to begin to assess another – broader – past; a past that includes philosophers, to be sure, as well as marchesas, entomologists, artists, hippologists and convicts. In addition, they force us to recognise that the worlds of these people exist only because of the presence of the beings with whom they interact: the sleeve cats and lap dogs; baited lions and inarticulate apes; wrestling bears and racing horses; feeling insects and dying geese. In recognising how assessments of the past have until now frequently failed to address the role of animals, and the significance of their interactions with us and with each other to human lives, history might begin to move forward in a new way.

Notes

1 Jacques Derrida, *The Animal That Therefore I Am*, ed. Marie-Louise Mallet, trans. David Wills (New York: Fordham University Press, 2008), 5.
2 Derrida, *The Animal*, 40–41.

Introduction

Action, reaction, interaction in historical animal studies

Sarah Cockram and Andrew Wells

> 25 April 1664. *[...] In the Duke's chamber there is a bird, given him by Mr. Pierce the surgeon, comes from the East Indys – black the greatest part, with the finest coller of white about the neck. But talks many things, and neyes like the horse and other things, the best almost that ever I heard bird in my life.*[1]

Samuel Pepys certainly loved animals. Like many well-to-do Englishmen of the 1660s, he managed to dine his way through a considerable number of species (oysters, swan, and venison were particular favourites), but his relationship with the animal world was more than merely gustatory. He kept horses, dogs, and cats, attended bear-baits and cock-fights, was a keen pigeon fancier, an inquisitive natural philosopher (who followed the latest developments in medicine, embryology, and microscopy), and was more than once afflicted by parasites and rodents. This range of interactions with animals was associated with a wide array of emotions – admiration of a horse-mimicking parrot, anxiety as a different bird almost blinded a companion, wonder at inspecting a louse under a microscope, irritation with his own lice infestation, fear upon hearing his cat bounding down stairs in the middle of the night, and so on – and extended beyond simple unidirectional engagement with animals to encompass a set of multilateral interactions, including between two or more different (species of) animals.[2]

These are the themes on which this book is focused, and a brief sketch of the range, diversity, and emotional implications of Pepys's interactions with dogs in particular will serve to illustrate the approach of this collection and expose some of the problems, challenges, and rewards of such a perspective. Pepys had a love-hate relationship with canines. He appreciated the beauty of certain breeds to the extent that he contemplated stealing a particularly handsome animal, and he was anxious for the wellbeing of a favourite dog. While on his way to Woolwich by water with his father, wife, and servant in April 1663, Pepys realised that they may have left their dog behind after he had followed the party to the bankside, 'which did not only strike my wife into a great passion but I must confess myself also; more than was becoming me'. Pepys would mourn the death of his dog Fancy in 1668 as 'one of my oldest acquaintances and servants'.[3] In addition to good service, dogs could provide amusement while even teaching political lessons: during Charles II's triumphant voyage from the Netherlands to England in May 1660, Pepys travelled with one of the king's favourite dogs, 'which shit in the boat, which made us laugh, and me think that a King and all that belong to him are but just as others are'.[4] But defecation was less edifying in his own household, where Pepys could become profoundly irritated with untrained dogs. During an argument with his wife Elizabeth in February 1660, Pepys threatened to defenestrate a newly acquired gift from his brother-in-law because it continued to urinate in the house: 'So to bed, where my wife and I had some high words upon my telling her that I would fling the dog which her brother gave her out at the window if he pissed the house anymore'.[5] The problem evidently lingered as in November Pepys locked the dog in the cellar, leading to another protracted argument that ruined his sleep.[6]

His rest had been interrupted earlier that year by the more direct influence of a neighbour's dog that barked throughout the night. Barking was one of the modes by which dogs interacted with Pepys, serving alternately to frighten and reassure him. While on the road in 1666 and feeling the urge to relieve himself, Pepys entered an empty inn and was in a vulnerable position when he heard the barking of a 'great dogg'; he

> was afeard how I should get safe back again, and therefore drew my sword and scabbard out of my belt to have ready in my hand, but did not need to use it, but got safe into the coach again, but lost my belt by the shift.

On the other hand, when Pepys was lost on the river at night in August 1662 and was anxiously moving up and down the Thames in an attempt to find his bearings, he suddenly realised that he was at Blackwall, thanks to the guidance provided by a barking dog he had seen earlier.[7]

Pepys also commented on and sought to facilitate interactions between animals. In September 1661, he visited the garden of Dr John Williams, who had trained his dog to kill and bury the cats that preyed on his pigeons. According to Williams, the dog performs this task

> with so much care that they shall be quite covered; that if but the tip of the tail hangs out he will take up the cat again, and dig the hole deeper. Which is very strange; and [Williams] tells me that he do believe that [the dog] hath killed above 100 cats.[8]

Pepys and his wife were more actively involved in creating than destroying life, but the size of the dogs involved brought problems. In early 1660, Elizabeth stayed up late to care for a parturient bitch as she was concerned that the dog was in danger from delivering what, from the size of the sire, were expected to be prodigiously large pups. A few years later, a handsome but small sire belonging to Mrs Buggins created the opposite problem: Pepys was particularly enamoured of this dog and wanted it to produce pups with his bitch, but it was so small that he was required to intervene. So

> I took him and the bitch into my closet below, and by holding down the bitch helped him to line her, which he did very stoutly, so as I hope it will take, for it is the prettiest dog that ever I saw.[9]

Pepys took the dogs to his closet because he was squeamish about animals having sex in front of his wife and servants. He thereby demonstrated that visibility and space are important elements in interspecies interactions, and excessive proximity – as much as the excessive distance manifested in their missing dog of April 1663 – could be exasperating. The lice and gnats that invaded Pepys's wig, covered his head and body, or plagued him in bed were only the most extreme example of uncomfortably shared space.[10] On 23 January 1669, itching provoked Pepys to have his wife de-louse him, and he pondered not only the number of lice that had managed to feed from him, but also the source of this infestation:

> So to my wife's chamber, and there supped, and got her cut my hair and look my shirt, for I have itched mightily these 6 or 7 days, and when all comes to all she finds that I am lousy, having found in my head and body about twenty lice, little and great, which I wonder at, being more than I have had I believe these 20 years. I did think I might have got them from the little boy, but they did presently look him, and found none – so how they come I know not; but presently did shift myself, and so shall be rid of them, and cut my hayre close to my head, and so with much content to bed.[11]

In September 1663, Pepys had similarly contemplated (as have many of us) why he was so appealing to biting insects when his companions were spared. Retiring to bed 'in a sad, cold, nasty chamber', he woke in the morning to find that he 'was bit cruelly, and nobody else of our company, which I wonder at, by the gnatts'.[12]

Pepys's musing over why he, over others, should be the target of seemingly not indiscriminate insect bites reminds us of the inscrutability of much animal agency. The essays in this volume seek to explore the full range of interactions between animal species in the early modern era – including parasitic ones – and encompass not only the mammal (in the home or the field, on show, on the plate, or in the wild), but also birds, fish, reptiles, and insects. Species encountered here range from useful domesticates and charismatic megafauna to those like Pepys's 'gnatts' who might be termed 'uncharismatic microfauna' and whose interaction with humans was to be avoided.[13]

Why interspecies interactions?

Why study these multiple interspecies interactions? If the role of the historian is to attempt to understand the human experience in the past and to consider what this might mean for us today, where does the animal fit in? Writing the animal into such studies has been done in various, often sensitive and illuminating, ways, as discussed below. Such studies can reveal the central place of the animal in human culture, economy, and society. One aim of this book is to join those studies that seek to push this enterprise further and write history that takes animal experience equally seriously. Animals are not only 'good to think [with]'[14] as a way to understand ourselves but as a way to understand animals and the relationships they have with humans and us with them.

Although the 'animal turn' is by now established, and historical animal studies is a growing field within a broader movement, the battle for its acceptance is not yet won. Reminiscent of the parody piece published in the *Journal of Social History* in 1974 and discussed by Erica Fudge in 'A Left-Handed Blow' (2002),[15] *The Guardian* newspaper in March 2016 reported on another satirical essay, this time published in the journal *Totalitarismus und Demokratie* (Totalitarianism and Democracy). The headline read: 'Editors of Dresden-based journal apologise after being fooled by fake PhD student's paper on role of alsatians in totalitarianism'.[16] According to online articles by the anonymous perpetrators, a hoax conference paper and subsequent journal publication were aimed at lampooning 'the latest academic fashion' of human–animal studies.[17] As the hoaxers see it, the 'animal turn', like other 'turns', reflects the academic compulsion 'which every few years must drive a new sow through the village'; continuing with the animal metaphor, the historical profession is later compared to a herd of cattle (*Rinderherde*).[18] The hoaxers attack a misplaced desire for radicalism, the development of a 'philosophical antihumanism', concepts of animal agency, and the conceptualisation of animals as 'the ultimate subaltern'.[19] This last criticism reflects anxiety that historical animal studies draws attention away from the experiences of those people whose histories still need to be written.[20] It is clear, however, that historical animal studies can illuminate not only animal lives but also those of the people whose worlds were touched by animals, including very often subaltern lives. Furthermore, work on attitudes about animals and ideas of animality in humans, or about human–animal boundaries, necessarily encompasses and advances study of those people who were classed as closer to the bestial.[21]

Writing about historical interspecies interactions throws into sharp relief a number of features of our disciplinary approaches, tools and sources, and responsibilities. The quest in humanistic scholarship to reach the interior life of another individual – for the historian, one who may be long dead – is rendered explicitly more difficult when that individual is a bat (as for Thomas Nagel) or a cow (as for Erica Fudge).[22] We must read our sources against the grain and with creativity, as exemplified in the chapters of this volume.

The ethical and political real-world dimensions of our choices as scholars are visible in historical animal studies. At our most ambitious, we attempt to engage seriously with animals, to say meaningful things about animals as both historical agents and as individuals and beings in the world. To write animal experience as well as animal significance back into the historical record is a call to respect for the natural world not only in our past but for our present and future.

The evolution of historical animal studies

Far from the fad targeted by the *Totalitarismus und Demokratie* hoaxers, scholarship on the history of humans and the animal world is now several decades old and has produced a sizeable body of literature, which has itself inspired a range of associated developments such as university courses, professional associations, symposia and conferences, including that which gave rise to this volume.[23] It is a field characterised by a wide array of different approaches and powerful interdisciplinary possibilities that extend beyond traditional collaborations (for example, between history, literature, anthropology, and archaeology) into realms of intellectual activity not usually touched by humanistic inquiry, such as animal psychology, veterinary practice, and zoology, to name but a few. Despite, or even because of this diversity and dynamism, it has not been a simple matter to categorise this field as a discrete area of scholarly activity. Contributors to such a body of scholarship have their own interests, goals, and intellectual and disciplinary backgrounds, all of which serve to make it appear perhaps bewilderingly heterogeneous.

The problem of defining this field, a necessary first step in assessing its contribution to the humanities and identifying potentially fruitful paths for development, was recognised by Fudge's 'Left-Handed Blow', where she characterised the subject as the 'history of animals', but did so, following Derrida's precept in his *Of Grammatology* (1967), 'under erasure'.[24] The 'history of animals'/'animal history' is, however, only one of several possible ways of denoting this body of scholarly work and is one, as Fudge rightly notes, hardly adequate to the task. For one thing, it promises a history that is hardly possible given the current state of methodological sophistication and the limitations of the sources from which we work.[25] A real history of animals must remain beyond our reach as long as we are only able to use sources created by humans. As László Bartosiewicz's chapter in this volume demonstrates, other (archaeological) evidence left by animals can help us to overcome some of the limitations of our sources, but this can only take us a few faltering steps towards a 'history of animals'.

What we are left with, therefore, is a history of human–animal relations that is told by only one side of that relationship. Fudge has outlined three modes of such a history, which she characterises as intellectual, humane, and holistic. Intellectual (which may perhaps be also termed 'cultural') histories deal with representations of animals by humans. They explore attitudes and ideas about animals in order to elucidate aspects of the broader cultures in which they appear. Those cultures are the ultimate object of such studies, amongst which, for the period covered by this book, may be numbered Keith Thomas's *Man and the Natural World: Changing Attitudes in England, 1500–1800* (1983), Linda Kalof's *Looking at Animals in Human History* (2007), Simona Cohen's *Animals as Disguised Symbols in Renaissance Art* (2008) and Bruce Boehrer's *Animal Characters: Nonhuman Beings in Early Modern Literature* (2010).

Humane studies move beyond the level of representation and symbolism to explore actual lived interactions between humans and animals. This might be within a quotidian domain but often relates to wider issues, especially of animal welfare, companionship, and the use of animals. Such studies exhibit a deep interest in the influence of human-animal interactions on the nature and development of human societies. Works like Virginia DeJohn Anderson's *Creatures of Empire* (2004) and collections such as Aubrey Manning and James Serpell's *Animals and Human Society* (1994), Anthony Podberscek, Elizabeth Paul, and James Serpell's *Companion Animals and Us* (2000), and Dorothee Brantz's *Beastly Natures* (2010) all fit this category.

Holistic studies take this one step further and examine what animals, and humans' relations with them, can mean for the category of the human. Erica Fudge and Harriet Ritvo are the pre-eminent scholars in this regard, but among other influential studies and collections is Lorraine Daston and Gregg Mitman's *Thinking with Animals* (2005), as well as numerous philosophical studies – particularly of Montaigne and Descartes – that approach these concerns without ever getting their hands dirty, feet muddy, or laps warmed by closely interacting with actual animals.[26]

All of these categories and the studies that populate them have at least one thing in common: the bilateral nature of interactions between humans and animals. Humans represent, use, slaughter, pet, yoke, and think with animals in this scholarship, but what is sometimes missing is attention to multidirectional interactions in which humans form but one or even no part. Of course, the problem of sources again rears its paralysing head, but it is one of the contentions of this book that a history of multilateral interspecies interactions is possible where we seek new evidence or read available evidence in innovative ways. To return to Pepys, this diary of an educated man obviously concentrates on his own activities and interactions with his environment and its human and animal inhabitants. But he does provide a glimpse of the ways in which early modern animals interacted with one another, from the savage and invariably fatal combats of cocks to, in a multispecies example, the dog trained by a human to kill and bury cats who prey on pigeons.[27]

To write intelligibly about interspecies interactions, we must become less anthropocentric (notwithstanding the well-taken point that human-centred perspectives are not *automatically* a bad thing) and conceive of our field of study not as human-animal relations, but as historical animal studies.[28] In so doing we open up to the fast-moving currents and stimulating conversation of cross-disciplinary animal studies more generally. Furthermore, we avoid the totalizing and unrealisable conceit of writing a 'history of animals', and we circumvent the limitations imposed by fixing one pole of our project squarely on humanity. With very few exceptions, all of our evidence derives from human beings, but new possibilities are opened up when these sources are read to place humans alongside and entwined with – not above – other animal actors (a contention enraging of course to our anonymous hoaxers).[29]

From an historical animal studies perspective, then, it becomes possible to interpret the body of scholarship that has shaped our field in a slightly different way. The categories of intellectual, humane, and holistic may be suggestively reconfigured as focusing on *actions*, *reactions*, and *interactions* (of course, a number of works have bridged and continue to bridge these overlapping domains). Intellectual and humane perspectives thus become those that respectively emphasise human *actions* alone or in conjunction with animal *reactions*. Humans have represented animals and scholars examine these representations to understand better the human cultures that produced them. Similarly, animals have reacted to human acts – as have humans to animal acts – and the interplay of these has played a substantial role in shaping human society.

This book seeks to contribute to a growing body of scholarship that concentrates more purely on *interactions*: humans and animals acting and reacting against, with, and between one another. Holistic studies approach this concept of interaction, but nevertheless remain focused on human views. Interspecies interactions, on the other hand, aim to decentre the human. A key concept in this approach is 'becoming with', recently discussed by Donna Haraway and others as a means of consciously and unconsciously shaping and refining human and animal identities. The following chapters explore how humans and animals shaped one another, how they produced emotions and acts in each other, and how they 'became with', bilaterally or more broadly (possibly excluding humans) through multispecies becomings.[30]

The essays in *Interspecies Interactions* explore these questions in what may be termed the long early modern period, between the fifteenth and nineteenth centuries, from the Renaissance to the mid-Victorian era. In these centuries of intellectual ferment, political upheaval, and dramatic social change, animals were a perennial topic of interest. Before the 'last century of the horse', their economic role was as pivotal as it had been since domestication, but human intellectual, social, cultural, political, and industrial revolutions forced a continual process of examination of animals and the human relationship with them.[31] Animals became and were to remain throughout these centuries 'good to think [with]'.

It is of course unwarranted to view this period as simply a march of progress towards modernity and greater animal welfare (even with the legal protections that came into force), nor can we chart a direct trajectory in attitudes to animals from Petrarch to Montaigne to Descartes to Bentham and beyond.[32] Nor should we forget that there was tremendous variety in attitudes to and relationships with and between animals before our period, as

Joyce Salisbury and others remind us.[33] But the early modern era was an unusually rich period for the development of concepts of the human and animal, as well as one in which the range of activities that humans and animals shared expanded remarkably. The following chapters convey something of the range and richness of interspecies interactions during this critical period and, it is hoped, encourage critical scrutiny of stereotypes of human-animal relations that rest on a wholeheartedly functionalist Middle Ages, a thoughtful but brutal Renaissance, an emphatically scientific Enlightenment, or a theriophilic Victorian age.

Approaching interaction

This volume brings together ten essays, in addition to fore- and afterwords, which reflect on the theme of interspecies interactions in the period between the Middle Ages and modernity. In Part One, 'Empathy, emotion, and companionship' are considered through three contributions. Ido Ben-Ami examines emotions towards animals elicited by an Ottoman festival celebrated in Istanbul in 1582 and analyses the representation of emotional responses to animals in the manuscript describing the event. Animal performers and the connection of these with famed creatures familiar to the book's audience prompt a range of feelings in the emotional community of the Ottoman elite. Sarah Cockram's essay similarly explores the emotional connections with animals in an elite environment, with her study of companion animals at the Italian Renaissance court of Mantua, and Pia Cuneo considers the possibility of giving early modern horses an autobiographical voice. In so doing, empathic possibilities are brought to life.

The second section of the book leads on from Cuneo's sometimes harrowing evidence concerning the mistreatment of horses by focusing on 'Use and abuse'. Like Cuneo's chapter, Peter Edwards' reconstructs stages in the life cycle of the horse, with a biography of the Levinz Colt (1721–1729), who was born and lived at Lord Edward Harley's stud at Welbeck, England. Although even the best-bred horses could end up as food for the hounds at the end of their lives, rarely would they end up on the early modern dinner plate, whence Karen Raber's chapter takes us. Zombie meats and concocted hybrid animals might be produced from the kitchen, in ways that complicated ingestion of another species and problematised the gustatory interactions of humans and animals. Andrew Wells's chapter stays with concepts of blurred boundaries in its analysis of the Enlightenment thought experiment that aimed to establish the borders of human and ape once and for all by attempting to breed a human with a primate. In the last essay of the section, Helen Cowie demonstrates the discomfort of many in nineteenth-century England at the idea of lion baiting, examining the opposition to the Warwick Lion Fight of 1825 and discussing the multiple interspecies interactions at its heart, with dogs and lions – as well as people – in conflict.

The third and final section of this volume raises questions of 'identification and classification' in its exploration of 'Self and other'. Dominik Hünniger takes up a metaphorical net and magnifying glass to capture and reveal the insects so prized by Enlightenment entomologists. László Bartosiewicz makes evident the inter-relationship between bovine and human in the construction, both of Hungarian Grey Cattle and Hungarian national identity, while Krista Maglen argues for the importance of dangerous native animals in the development of an Australian settler identity as well as the policing of early colonial settlements.

In addition to the substantial overlap between the themes around which the three sections of this book are organised, its chapters speak to each other in a number of ways that serve to highlight the challenges and opportunities of historical animal studies highlighted above. A range of conceptual underpinnings and methodologies is deployed by contributors. The majority draw evidence from archival sources and documents of various kinds, including recipe books (Raber), legal records (Wells), letters (Cockram), works of natural history (Wells, Hünniger), and newspapers (Cowie). Several make use of literature (Cuneo, Maglen), art (Cuneo, Cockram), and archaeology (Bartosiewicz, Cuneo). Modern research from beyond the humanities is drawn on by Cockram and Wells. Current debate about the utility – and possibility – of animal biography and the reconstruction of individual animal lives is taken up by Cuneo and Edwards.[34] The enduring vitality and urgency of questions about human-animal boundaries and anthropomorphism are manifest in most contributions.[35] While some chapters are about individuals, others are about groups or larger collectives at the level of nations or breeds (Bartosiewicz). The influence of the breeding of domesticates on human identity, aspiration, and feelings is a recurring theme across several chapters (Cockram, Bartosiewicz, Raber, Cuneo, Edwards).

We began this introduction by pointing out the range of animal species in this book. The range within the species of *homo sapiens* might also be considered, with attendant questions of gender, power, and status. We meet, among others, people who work closely with animals, dead or alive (cooks, tanners, farriers, artists, entomologists, drovers, grooms, menagerists, trainers of dancing bears); dealers in animals; audiences of animal spectacles; and convicts eaten by sharks. The geographical habitats of the animals and humans stretches across Europe, the Near East, Africa, and Australia. We regret that the confines of the volume have not made it possible to include anything more than passing references to the Americas and Far East, two locations ripe for further study.[36]

One call that we make for the future of historical animal studies, then, and echoing Sandra Swart, is for work that continues to develop non-Euro or Anglocentric topics.[37] As historical animal studies sheds its status as a marginal or quirky scholarly occupation, we continue, in the image conjured up by our anonymous German detractors, 'to drive a new sow through the village', adding to our understanding of historical interspecies interactions as we go. Or, while we can still be viewed as radical (and the day that we are not will, we hope, augur better days for the animals with whom we live), let us return to Samuel Pepys for our model and be the nit on the wig that subverts the respectable image or the little dog that messily disturbs the decorum of the house, hoping that no one will throw us out of the window.

Notes

1   *The Diary of Samuel Pepys*, ed. Robert Latham and William Matthews, 11 vols (London: Bell & Hyman, 1970-83), 25 April 1664.

2   Pepys, *Diary*, (food) 19 January 1663, 2 March 1663, 2 September 1663, 12 January 1664, (cat) 31 December 1660, 29 November 1667, (dog) 17 February 1664, (bear-pit) 9 September 1667, (cock-fight) 8 September 1666, (pigeon-fancying) 8 February 1660, (medicine: blood transfusions) 14 November 1666, (embryology: spontaneous insect generation) 23 May 1661, (microscopy) 13 February 1664, (parasites) 8–13 July 1661, (rodents) 31 December 1660, (parrot) 19 March 1665, (lice) 13 February 1664, 23 January 1669.

3   Pepys, *Diary*, 8 April 1660, 16 September 1668.

4   Pepys, *Diary*, 25 May 1660.

5   Pepys, *Diary*, 12 February 1660.

6   Pepys, *Diary*, 7 August 1661, 8 April 1663, 12 February 1660, 6 November 1660.

7   Pepys, *Diary*, 15 January 1660, 28 January 1666, 4 April 1662.

8   Pepys, *Diary*, 11 September 1661.

9   Pepys, *Diary*, 22–23 March 1664.

10   Pepys, *Diary*, (lice and gnats) 17 September 1663, 19 September 1663, 3 September 1664, 23 January 1669, (infested wigs) 27 March 1667, 4 April 1667.

11   Pepys, *Diary*, 23 January 1669.

12   Pepys, *Diary*, 17 September 1663.

13   See Dominik Hünniger's chapter in this volume for more on uncharismatic microfauna.

14   Claude Lévi-Strauss, *Totemism*, tr. Rodney Needham (London: Merlin, 1991 [1962]), 89.

15   Charles Phineas, 'Household Pets and Urban Alienation', *Journal of Social History*, 7 (1973/4), 338–43; Erica Fudge, 'A Left-Handed Blow: Writing the History of Animals', in Nigel Rothfels (ed.), *Representing Animals* (Bloomington: Indiana University Press, 2002), 3–18 at 4–5.

16   'Human-animal studies academics dogged by German hoaxers', *The Guardian* (online), 1 March 2016, www.theguardian.com/world/2016/mar/01/human-animal-studies-academics-dogged-by-german-hoaxers, accessed 7 July 2016.

17   'der neuesten akademischen Mode'. Christiane Schulte (pseud.) and Friends [*Freund_innen*], 'Kommissar Rex an der Mauer erschossen?', *Telepolis* (online), 15 February 2016, www.heise.de/tp/artikel/47/47395/1.html, accessed 7 July 2016.

18   'Gleichzeitig bedienen sie den akademischen Profilierungszwang, der alle paar Jahre eine neue Sau durchs Dorf treiben muss: Nach dem *cultural turn*, dem *linguistic turn*, dem *spatial turn*, dem *iconic/visual turn*, dem *body turn* und dem *emotional turn* ist jetzt also der animal turn an der Reihe'. Schulte and Friends, 'Kommissar Rex', www.heise.de/tp/artikel/47/47395/2.html, accessed 7 July 2016.

19   'eines philosophischen Antihumanismus', 'das Tier als das ultimativ Subalterne'. Schulte and Friends, 'Kommissar Max', www.heise.de/tp/artikel/47/47395/2. html and www.heise.de/tp/artikel/47/47395/3.html, accessed 7 July 2016. The hoaxers go on to state: 'While saying nothing against work that investigates either animals or the interaction of humans and animals or denounces cruelty

to animals, the problem is rather the radical anti-humanism of Human Animal Studies [...] Once the separation of natural and cultural history is missing, then the racism of the Atlantic slave trade is no longer distinguishable from the 'speciesism' of a pig farm'. (*Zwar spricht nichts dagegen, Tiere oder die Interaktion von Mensch und Tier zu untersuchen und Tierquälerei anzuprangern. Das Problem ist jedoch der radikale Antihumanismus der Human Animal Studies [...] Sobald die Trennung von Natur- und Kulturgeschichte entfällt, ist der Rassismus des atlantischen Sklavenhandels nicht mehr unterscheidbar vom "Speziesismus" eines Schweinemastbetriebes*).

20 Concern about excessive human affection for animals is also a theme in Midas Dekkers, *On Bestiality*, tr. Paul Vincent (London: Verso, 2000).

21 This is true both for literary and historical studies of human-animal relationships. For the former, see Philip Armstrong, 'The Postcolonial Animal', *Society & Animals*, 10 (2002), 414–19; Graham Huggan and Helen Tiffin, *Postcolonial Ecocriticism* (London: Routledge, 2010); Elizabeth DeLoughrey and George Handley (eds), *Postcolonial Ecologies* (Oxford: Oxford University Press, 2011). For more explicitly historical studies, see among others Marjorie Spiegel, *The Dreaded Comparison: Human and Animal Slavery* (London: Heretic, 1998 [expanded edn New York: Mirror, 1997]); Jennifer Wolch and Jody Emel (eds), *Animal Geographies: Place, Politics, and Identity in the Nature-Culture Borderlands* (London: Verso, 1998); Virginia DeJohn Anderson, *Creatures of Empire* (New York: Oxford University Press, 2004); Thomas G. Andrews, 'Beasts of the Southern Wild: Slaveholders, Slaves, and Other Animals in Charles Ball's *Slavery in the United States*', in Marguerite S. Shaffer and Phoebe S. K. Young (eds), *Rendering Nature: Bodies, Places, and Politics* (Philadelphia: University of Pennsylvania Press, 2015), 21–47.

22 Erica Fudge, 'Milking Other Men's Beasts', *History and Theory*, 52/4 (2013), 13–28 at 23–25; Thomas Nagel, 'What Is It Like to Be a Bat?', *Philosophical Review*, 83 (1974), 435–50.

23 Lists of courses offered and conferences on human-animal studies worldwide can be found on the Animals and Society Institute website animalsandsociety.org, at (respectively) www.animalsandsociety.org/human-animal-studies/courses and www.animalsandsociety.org/human-animal-studies/calls-for-abstracts. See also the 'Animal Studies Living Bibliography' at www.lbanimalstudies.org.uk. All websites accessed 7 October 2016.

24 Fudge, 'Left-Handed Blow', 7; Jacques Derrida, *Of Grammatology*, tr. Gayatri Chakravorty Spivak (Baltimore: Johns Hopkins University Press, 1997), chapter 1.

25 See the consideration of these issues in Sarah Cockram's chapter in this volume.

26 See, in particular, Harriet Ritvo, *The Animal Estate* (Cambridge: Harvard University Press, 1987); id., *The Platypus and the Mermaid and Other Figments of the Classifying Imagination* (Cambridge: Harvard University Press, 1997); id., *Noble Cows and Hybrid Zebras: Essays on Animals and History* (Charlottesville: University of Virginia Press, 2010); Erica Fudge, Ruth Gilbert and Susan Wiseman (eds), *At the Borders of the Human: Beasts, Bodies and Natural Philosophy in the Early Modern Period* (Basingstoke: Macmillan, 1999); Erica Fudge, *Perceiving Animals: Humans and Beasts in Early Modern English Culture* (Basingstoke: Macmillan, 2000); id. (ed.), *Renaissance Beasts* (Urbana: University of Illinois Press, 2004); id., *Brutal Reasoning: Animals, Rationality, and Humanity in Early Modern England* (Ithaca: Cornell University Press, 2006).

27 Pepys, *Diary*, 11 September 1661, 21 December 1663, 6 April 1668.

28 For commentators such as Mary Midgley, Tim Hayward, and Matthieu Ricard, anthropocentrism (in the sense of humans being the centre of their own moral universe) and anthropomorphism play a role in enabling the ethical treatment of animals by, respectively, producing moral judgements about human experience and then by recognising an equivalence of human and animal experience, in order

to apply those ethical judgements to animals. See Mary Midgley, 'The End of Anthropocentrism?', *Royal Institute of Philosophy Supplements*, 36 (1994), 103–12; Tim Hayward, 'Anthropocentrism: A Misunderstood Problem', *Environmental Values*, 6 (1997), 49–63; Matthieu Ricard, *A Plea for the Animals*, tr. Sherab Chödzin Kohn (Boulder, CO: Shambhala, 2016), esp. 131–36. On anthropocentrism in general, see Gary Steiner, *Anthropocentrism and Its Discontents* (Pittsburgh: University of Pittsburgh Press, 2005) and Rob Boddice (ed.), *Anthropocentrism: Humans, Animals, Environments* (Leiden: Brill, 2011).

29 On 'entwinement', see Erica Fudge's foreword to this volume.

30 See esp. Donna J. Haraway, *When Species Meet* (Minneapolis: University of Minnesota Press, 2008). We thank the anonymous reader for this book who saw multispecies becomings as key to this collection.

31 See Ulrich Raulff, *Das letzte Jahrhundert der Pferde* (Munich: C. H. Beck, 2015).

32 A selection of relevant literature includes: Benjamin Arbel, 'The Renaissance Transformation of Animal Meaning: From Petrarch to Montaigne', in Linda Kalof and Georgina M. Montgomery (eds), *Making Animal Meaning* (East Lansing: Michigan State University Press, 2011), 59–80; Andreas-Holger Maehle, 'Cruelty and Kindness to the "Brute Creation": Stability and Change in the Ethics of the Man-Animal Relationship, 1600–1850', in Aubrey Manning and James Serpell (eds), *Animals and Human Society: Changing Perspectives* (London: Routledge, 1994), 81–105; Angus Taylor, *Animals and Ethics: An Overview of the Philosophical Debate* (3rd ed., Peterborough, ON: Broadview, 2009), chapter 2; Peter Harrison, 'The Virtues of Animals in 17th-Century Thought', *Journal of the History of Ideas*, 59 (1998), 463–84.

33 Joyce E. Salisbury, *The Beast Within: Animals in the Middle Ages* (New York: Routledge, 1994); Brigitte E. Resi (ed.), *A Cultural History of Animals*, ii: *The Medieval Age* (Oxford: Berg, 2007); Susan Crane, *Animal Encounters: Contacts and Concepts in Medieval Britain* (Philadelphia: University of Pennsylvania Press, 2013).

34 Erica Fudge, 'Writing the Life of Animals', *History Today*, 54/10 (2004), 21–26.

35 Lorraine Daston and Gregg Mitman (eds), *Thinking with Animals: New Perspectives on Anthropomorphism* (New York: Columbia University, 2005).

36 As consolation for necessarily incomplete coverage here, we point the reader towards, as a starting point, Sandra Swart, *Riding High: Horses, Humans and History in South Africa* (Johannesburg: Wits University Press, 2010); Miguel de Asúa and Roger French, *A New World of Animals: Early Modern Europeans on the Creatures of Iberian America* (Aldershot: Ashgate, 2005); Aaron Herald Skabelund, *Empire of Dogs: Canines, Japan, and the Making of the Modern World* (Ithaca: Cornell University Press, 2011) and Anderson, *Creatures of Empire*.

37 Sandra Swart, review of Dorothee Brantz (ed.), *Beastly Natures: Animals, Humans, and the Study of History* (Charlottesville: University of Virginia Press, 2010), *H-Environment* (November 2011), https://networks.h-net.org/node/19397/reviews/20598/swart-brantz-beastly-natures-animals-humans-and-study-history, accessed 7 October 2016.

Bibliography

Anderson, Virginia DeJohn, *Creatures of Empire* (New York: Oxford University Press, 2004).

Andrews, Thomas G., 'Beasts of the Southern Wild: Slaveholders, Slaves, and Other Animals in Charles Ball's *Slavery in the United States*', in Marguerite S. Shaffer and Phoebe S. K. Young (eds), *Rendering Nature: Bodies, Places, and Politics* (Philadelphia: University of Pennsylvania Press, 2015), 21–47.

Arbel, Benjamin, 'The Renaissance Transformation of Animal Meaning: From Petrarch to Montaigne', in Linda Kalof and Georgina M. Montgomery (eds), *Making Animal Meaning* (East Lansing: Michigan State University Press, 2011), 59–80.

Armstrong, Philip, 'The Postcolonial Animal', *Society & Animals*, 10 (2002), 414–19.

Asúa, Miguel de and Roger French, *A New World of Animals: Early Modern Europeans on the Creatures of Iberian America* (Aldershot: Ashgate, 2005).

Boddice, Rob (ed.), *Anthropocentrism: Humans, Animals, Environments* (Leiden: Brill, 2011).

Boehrer, Bruce, *Animal Characters: Nonhuman Beings in Early Modern Literature* (Philadelphia: University of Pennsylvania Press, 2010).

Brantz, Dorothee (ed.), *Beastly Natures: Animals, Humans, and the Study of History* (Charlottesville: University of Virginia Press, 2010).

Cohen, Simona, *Animals as Disguised Symbols in Renaissance Art* (2008).

Crane, Susan, *Animal Encounters: Contacts and Concepts in Medieval Britain* (Philadelphia: University of Pennsylvania Press, 2013).

Daston, Lorraine and Gregg Mitman (eds), *Thinking with Animals: New Perspectives on Anthropomorphism* (New York: Columbia University, 2005).

Dekkers, Midas, *On Bestiality*, tr. Paul Vincent (London: Verso, 2000).

DeLoughrey, Elizabeth and George Handley (eds), *Postcolonial Ecologies* (Oxford: Oxford University Press, 2011).

Derrida, Jacques, *Of Grammatology*, tr. Gayatri Chakravorty Spivak (Baltimore: Johns Hopkins University Press, 1997).

Erica, Fudge, *Perceiving Animals: Humans and Beasts in Early Modern English Culture* (Basingstoke: Macmillan, 2000).

Erica, Fudge, 'A Left-Handed Blow: Writing the History of Animals', in Nigel Rothfels (ed.), *Representing Animals* (Bloomington: Indiana University Press, 2002), 3–18.

Erica, Fudge (ed.), *Renaissance Beasts* (Urbana: University of Illinois Press, 2004).

Erica, Fudge, 'Writing the Life of Animals', *History Today*, 54/10 (2004), 21–6.

Erica, Fudge, *Brutal Reasoning: Animals, Rationality, and Humanity in Early Modern England* (Ithaca, NY: Cornell University Press, 2006).

Erica, Fudge, 'Milking Other Men's Beasts', *History and Theory*, 52/4 (2013), 13–28.

Erica, Fudge, Ruth Gilbert and Susan Wiseman (eds), *At the Borders of the Human: Beasts, Bodies and Natural Philosophy in the Early Modern Period* (Basingstoke: Macmillan, 1999).

Haraway, Donna J., *When Species Meet* (Minneapolis: University of Minnesota Press, 2008).

Harrison, Peter, 'The Virtues of Animals in 17th-Century Thought', *Journal of the History of Ideas*, 59 (1998), 463–84.

Hayward, Tim, 'Anthropocentrism: A Misunderstood Problem', *Environmental Values*, 6 (1997), 49–63.

Huggan, Graham and Helen Tiffin, *Postcolonial Ecocriticism* (London: Routledge, 2010).

Kalof, Linda, *Looking at Animals in Human History* (London: Reaktion, 2007).

Kalof, Linda and Georgina M. Montgomery (eds), *Making Animal Meaning* (East Lansing: Michigan State University Press, 2011).

Lévi-Strauss, Claude, *Totemism*, tr. Rodney Needham (London: Merlin, 1991 [1962]).

Maehle, Andreas-Holger, 'Cruelty and Kindness to the "Brute Creation": Stability and Change in the Ethics of the Man-Animal Relationship, 1600–1850', in Aubrey Manning and James Serpell (eds), *Animals and Human Society: Changing Perspectives* (London: Routledge, 1994), 81–105.

Manning, Aubrey and James Serpell (eds), *Animals and Human Society: Changing Perspectives* (London: Routledge, 1994).

Midgley, Mary, 'The End of Anthropocentrism?', *Royal Institute of Philosophy Supplements*, 36 (1994), 103–12.

Nagel, Thomas, 'What Is It Like to Be a Bat?', *Philosophical Review*, 83 (1974), 435–50.

Pepys, Samuel, *The Diary of Samuel Pepys*, ed. Robert Latham and William Matthews, 11 vols (London: Bell & Hyman, 1970–83).

Phineas, Charles, 'Household Pets and Urban Alienation', *Journal of Social History*, 7 (1973/4), 338–43.

Podberscek, Anthony, Elizabeth Paul, and James Serpell, *Companion Animals and Us* (Cambridge: Cambridge University Press, 2000).

Raulff, Ulrich, *Das letzte Jahrhundert der Pferde* (Munich: C. H. Beck, 2015).

Resi, Brigitte E. (ed.), *A Cultural History of Animals*, ii: *The Medieval Age* (Oxford: Berg, 2007).

Ricard, Matthieu, *A Plea for the Animals*, tr. Sherab Chödzin Kohn (Boulder, CO: Shambhala, 2016).

Ritvo, Harriet, *The Animal Estate* (Cambridge: Harvard University Press, 1987).

Ritvo, Harriet, *The Platypus and the Mermaid and Other Figments of the Classifying Imagination* (Cambridge: Harvard University Press, 1997).

Ritvo, Harriet, *Noble Cows and Hybrid Zebras: Essays on Animals and History* (Charlottesville: University of Virginia Press, 2010).

Rothfels, Nigel (ed.), *Representing Animals* (Bloomington: Indiana University Press, 2002).

Salisbury, Joyce E., *The Beast Within: Animals in the Middle Ages* (New York: Routledge, 1994).

Shaffer, Marguerite S. and Phoebe S. K. Young (eds), *Rendering Nature: Bodies, Places, and Politics* (Philadelphia: University of Pennsylvania Press, 2015).

Skabelund, Aaron Herald, *Empire of Dogs: Canines, Japan, and the Making of the Modern World* (Ithaca: Cornell University Press, 2011).

Spiegel, Marjorie, *The Dreaded Comparison: Human and Animal Slavery* (London: Heretic, 1998 [expanded ed., New York: Mirror, 1997]).

Steiner, Gary, *Anthropocentrism and Its Discontents* (Pittsburgh: University of Pittsburgh Press, 2005).

Swart, Sandra, *Riding High: Horses, Humans and History in South Africa* (Johannesburg: Wits University Press, 2010).

Taylor, Angus, *Animals and Ethics: An Overview of the Philosophical Debate* (3rd ed., Peterborough, ON: Broadview, 2009).

Thomas, Keith, *Man and the Natural World: Changing Attitudes in England, 1500–1800* (London: Penguin, 1983).

Wolch, Jennifer and Jody Emel (eds), *Animal Geographies: Place, Politics, and Identity in the Nature-Culture Borderlands* (London: Verso, 1998).

Periodicals and websites

Animal Studies Living Bibliography www.lbanimalstudies.org.uk.

Animals and Society Institute www.animalsandsociety.org.

H-Environment networks.h-net.org/h-environment.

Telepolis (online).

The Guardian (online).

Part I

Empathy, emotion, and companionship

1 Emotions and the sixteenth-century Ottoman carnival of animals

Ido Ben-Ami

Engaging animals with Ottoman emotions

In this chapter, I will explore some aspects of the emotional engagement of Ottoman elites with animals in sixteenth-century Istanbul. To do so, I will analyse various emotional responses directed towards animals as depicted in *The Imperial Book of Festival* (*Surname-i Hümayun*), a manuscript text dedicated to the description of a royal Ottoman festival held in Istanbul in 1582.[1]

Traditionally, the Ottomans marked special political occasions by throwing lavish festivals.[2] In 1582, such a celebration was held in honour of the circumcision of Sultan Murad III's (r. 1574–1595) sixteen-year-old son, Prince Mehmet, later to become Sultan Mehmet III (r. 1595–1603). The festival lasted for more than fifty days and nights while its main events took place at the ancient Byzantine Hippodrome (*Atmeydanı*) in Istanbul. The official programme of the event included formal receptions, an exchange of gifts, ostentatious feasts and firework displays. In addition, a variety of entertainments given by the Empire's most talented artists demonstrated their skills for the pleasure of their audience. Among them there were rope dancers, acrobats, tumblers, jesters, jugglers, wrestlers, musicians, jockeys and animal trainers who interacted with exotic as well as domesticated animals, to the amazement of their audience.[3]

In the early modern period, when zoos in their modern form did not exist, this kind of animal show gave urban Ottomans a unique opportunity to encounter different types of fauna. Due to the fact that such performances were staged by skilful trainers who taught creatures to perform according to well-planned theatrical acts, these encounters did not always introduce animals on their own terms, but rather in the guise of humans or even as creatures that surpass human intelligence.

At these shows, special emotional atmospheres were created by the performances of animals and their trainers. This was especially discernible during performances that included predators, during which their trainers created an emotional atmosphere of joy instead of the more usual fear that spontaneous and uncontrolled encounters with these animals might be expected to produce. Relying on the Bakhtinian definition of the carnivalesque, whereby an atmosphere based upon humour and laughter exists, I will argue that this special sense prevailing during the Ottoman festival had an impact on emotional responses elicited by the performing animals.[4]

The assumption that the generation of emotion towards animals depends on social and cultural practices such as carnival might seem straightforward, but it raises an important observation concerning the nature of feelings in general: the claim that emotion is influenced by external factors is a point of view that follows social construction models. It assumes that feelings are products that are prescribed 'by the social world and constructed by people, rather than by nature'.[5] This approach aims to challenge other models of the generation of emotion known as 'basic emotion models'.[6] According to these models, feelings are biologically basic states caused by a mechanism that is not influenced by external elements such as culture.[7] While the latter models are widely accepted among neuropsychologists, scholars in the humanities and historians in particular may find them inapplicable because they convey the idea that emotions are 'hardwired' and thus cannot change over time.[8]

Examining how emotions were constituted by sociocultural factors allows us to focus on a certain social group whose members are part of a unique emotional community. As pointed out by Barbara Rosenwein, this kind of community shares the same norms, values and beliefs that determine how its members think, act and feel on certain occasions.[9] These communities can be in the form of 'families, neighbourhoods, parliaments, guilds, monasteries, parish church memberships – but the researcher looking at them seeks above all to uncover systems of feeling'.[10] Of course, every given society contains more than one emotional community, and several – perhaps all – such groups frequently overlap with each other.[11]

I will thus consider the Ottoman elite of the sixteenth century as such an emotional community that encompassed several different subcommunities: the imperial family and court, alongside bureaucrats, artists, historians and others. As pointed out by Walter Andrews, this community was made up of powerful, educated individuals who produced cultural scenarios (stories, poems, images, etc.) that contained an emotional vocabulary (words, images, music and symbols, as well as other cultural artefacts).[12] As Andrews points out, 'when significant responses to various stimuli are captured in an emotional vocabulary, the result is the emotional content of social interactions'.[13] That is to say that by means of these scenarios, emotions were expressed and understood by the Ottoman elite in this period.

Therefore, in addition to reading *The Imperial Book of Festival* to analyse face-to-face encounters with animals during the performances, I also read it with the intention of uncovering the themes and emotional vocabulary that those members of the Ottoman elite highlighted as commendable for their social milieu regarding performing animals. I will argue that the use of such vocabulary indicates that the book was designed by such individuals to re-create the emotional atmosphere of the Ottoman carnival and cast these as unique emotional experiences for its elite readership. Such re-creation allows us to grasp how this group defined its relations with animals as well as to observe how it shaped its uniqueness as an emotional community.

Before I discuss the Ottoman carnival of animals in detail, I will examine one performance of the festival – that of a lion – in order to explain why I prefer the social construction model of emotions, as well as to give some essential background on the sponsors of *The Imperial Book of Festival*.

Constructing glorification of the sultan through the use of Ottoman awe

One of the performances during the festival of 1582 was meant to introduce a staged battle between a lion and a boar.[14] The author of *The Imperial Book of Festival*, İntizami (the 'Organizer'), describes a most 'dreadful' lion that entered the Hippodrome. He recalls that this animal theoretically 'could have shaken the constellation of Aries with a single stroke and caused Leo to fall down lamenting'. Additionally, 'if an elephant had come into his possession, this lion could have torn this animal into pieces with his paw'.[15]

However, something apparently went wrong during the performance since the lion refused to perform. According to İntizami, although 'this animal was awed by [the presence] of the sultan, he wanted to go back to his [lion]-house'.[16] As part of the show, the lion keepers brought to the field a vigorous boar in order to hold a fight between the two animals, but the lion did not pay attention to this animal as he was looking to the other side.

Two different emotional reactions feature in İntizami's descriptions regarding this lion: human fear of this wild cat as well as the awe it supposedly felt at the presence of the sultan. Fear of wild animals might be considered as an instinctive universal reaction: as Paul A. Trout suggests, evolution helped both humans and animals to react fearfully to a general set of physical stimuli that became associated with dangerous animals (staring eyes, big mouth, pointed teeth, etc.).[17] Thus, the perception of these stimuli and not the creature itself elicits fear.

Furthermore, Trout mentions that these stimuli were used by Paleolithic storytellers 'as "stage props" to enact "tales" about predators'.[18] Thus, each society uses its own stories in which animals are viewed or treated.[19] Although such stories are not evolutionarily adaptive, their repetition by a certain emotional community helps to stimulate emotions towards animals among this group according to social construction models of emotion generation.

Therefore, by mentioning the frightening stories regarding the lion and the zodiac, İntizami tried to stimulate fear. However, during the Ottoman festival, the trainers sought to use the fearsomeness of lions to create a sensation of awe for Sultan Murad. This was to be achieved through the staged battle between the lion and the boar. As noted by Nurhan Atasoy, the battle was understood by viewers who attended the festival as a staged symbolic combat between Islam and Christianity: if the lion won, it may have been regarded as a joyous occasion, but since he lost, the whole incident may be regarded as an evil omen not only for the victory of Islam, but also for Sultan Murad as a legitimate ruler.[20]

Within *The Imperial Book of Festival*, the risk that the animal would fail and thus not fulfil its symbolic mission could not be tolerated. Therefore, İntizami reinterpreted the event for his readers by constructing his narrative in such a way as to blunt the symbolic import of the lion's defeat. Thus, according to the author, although the lion was awed by the presence of the sultan, his disobedience was actually intended. İntizami explains that this was due to the fact that the lion's trainers 'spoke the language of the Melâmeti [and thus] kept doing their crafts [deliberately] with deficiency'.[21] By suggesting that the trainers belonged to the Melâmetiye, a sect of Sunni dervishes who subjected themselves to public reproach, İntizami managed to defend Sultan Murad's image as awe inspiring since the lion's failure to perform his symbolic mission was intentional.[22]

As pointed out by Dacher Keltner and Jonathan Haidt, an attempt to depict a charismatic leader by using an emotional description of awe takes place when there is an interest to stress power dynamics and the maintenance of the social order.[23] Following this notion, the fact that İntizami made an effort to stress awe from the performing lion even when it failed to perform suggests that his emotional community had a collective interest in reading the incident with this feeling. Indeed, within several manuscripts produced by the Ottoman elite during the early modern period, awe was stressed with the use of various animals. This was especially noticeable within hunting descriptions in which the sultans were portrayed as pursuing dangerous creatures.[24]

However, İntizami's description of awe was different from these hunting scenes. The motivation for depicting the lion's performance in this manner lies in *The Imperial Book of Festival*'s own history, as well as its sponsors and their political and social agendas. Since such events as the festival of 1582 carried great political significance, demonstrating the continuity of the Ottoman dynasty as well as its grandeur and glory, considerable efforts were invested in recording these events in writing, which produced the literary genre of the books of festivals.[25]

İntizami's book, which was completed six years after the festival itself took place, was commissioned by Sultan Murad. This sultan was known as an enthusiastic bibliophile who dedicated much of his attention to expanding the Ottoman collection of manuscripts.[26] Two of the sultan's companions supervised the making of this book: Mehmet Agha, the chief black eunuch of the imperial harem and his colleague Zeyrek Agha. According to İntizami's own records, he consulted both of them as they gave his work both moral and material support.[27]

It was unusual that eunuchs of such lowly status would sponsor *The Imperial Book of Festival*. Other books were written by far more significant members of the Ottoman elite. According to Emine Fetvacı, the shift in sponsorship indicates important changes in the Ottoman social hierarchy.[28] Fetvacı suggests that the book records the festival as befitted the personal points of view of these two eunuchs. Thus, they picked İntizami, who was merely a scribe and not a professional historian of the Ottoman court.[29] Consequently, the image of Murad was constructed by İntizami using different characteristics to those commonly attributed to Ottoman rulers. This sultan was a sedentary ruler and thus his legitimacy to rule is different: within *The Imperial Book of Festival* Murad's warfare or hunting skills are not discussed, but rather he is cast as the vice-regent of God on Earth.[30]

By keeping in mind the aim of portraying Sultan Murad in this manner, we can better understand İntizami's attempt to depict him as the ultimate ruler of the animal kingdom, by whom even its king, the lion, would be awed. Associating lions with this emotional sensation suggests the way that this animal was perceived by the Ottoman elite. As Alan Mikhail suggests, the Ottoman conquest of Egypt in 1517 established an Indian Ocean trading network that enabled exotic animals such as lions to move easily into and out of the Ottoman Empire. This network facilitated the use of charismatic megafauna such as lions in order to emphasise that the Ottoman rulers could control nature. Additionally, lions were usually brought as gifts to the sultans from the Empire's North African provinces or from Safavid Persia.[31] During the sixteenth and seventeenth centuries in Istanbul, these animals were well cared for and kept at the Imperial Lion House known as the 'Arslanhane'.[32]

Lions were only part of the Ottoman festival of 1582; there were many other animals and performances in which animals played significant roles. These various shows were not only for constructing the image of the sultan in the eyes of the author (or commissioners of texts) but also a window into the emotional world of early modern Ottomans towards the animal kingdom.

An emotional Ottoman carnival of animals

Many of the performances during the Ottoman festival are characterised by the 'world upside-down' principle of carnivals. As Bakhtin noted in *Rabelais and His World,* carnival festivities occupied an important place in medieval and early modern life. During these special occasions, people built themselves an alternative world, one completely out of the ordinary. This world was based on characteristics of laughter that, according to Bakhtin, was 'determined by the traditions of the medieval culture of folk humour'.[33] Hence, while official feasts were used to assert all that was stable in medieval life, carnivals offered a temporary liberation from hierarchy, norms, religion and even prohibitions. Even though this alternative world did not exist during official feasts sponsored by the church, official medieval festivities still had what Bakhtin recognised as 'a carnival atmosphere' that included 'varied open-air amusements, with the participation of giants, dwarfs, monsters, and trained animals'.[34]

Ottoman festivals were certainly meant to be seen as official royal celebrations rather than '*festa stultorum*'. Therefore, they did not try to defy the sultan's place at the top of the social hierarchy. As pointed out by Fariba Zarinebaf, the playful inversion of sociopolitical hierarchies occurred in Ottoman festivals 'without going too far' especially within the official descriptions of the festivals in imperial books.[35] Nevertheless, since these festivals still had some elements of carnival, Derin Terzioğlu suggests in her interpretation of the festival of 1582, that it 'was somewhere between an official feast and a carnival'.[36]

The carnival elements during Ottoman festivals were thus emphasised mostly through performances in which this alternative world was introduced to spectators for a strictly limited time. As Shulamith Lev-Aladgem has suggested, Bakhtinian carnival performances occur during a special mode of acting classified as 'carnivalesque enactment'. During this liminal 'betwixt and between' mode, performers are allowed to engage in absurd dramatic play-acting, which allows them to experience intense fun and humour. This was therefore a special situation that 'did not seriously intend to construct a complete alternative for the social order'.[37]

Spectators of Ottoman festivals were also aware of this special temporary mode during animal performances: for example, in 1720 another Ottoman festival took place in Istanbul and was recorded within an imperial book.[38] In one part, the author Vehbi (1674–1736) describes a staged wrestling contest between a man called Çomar ('mastiff') and a bear. He relates that the spectators of this frightful performance were 'show experts' who understood the situation that unfolded in front of their eyes and thus 'became amused and laughed with each other'.[39] In other words, the author explains here how the mode of carnivalesque enactment was recognised by the Ottoman audience that was aware of the fact that it was a just a show.

The idea of carnivalesque enactment was also realised through several animal shows in the festival of 1582: during one performance, İntizami describes how a cat complied with a 'secretive sign' (*açmazdan bir işâret*) from its trainer, took a training stick in its paws and began to juggle over a cord while demonstrating skills such as passing through hoops and dancing gracefully in the Arab and Persian styles. At some point, however, the cat ceased its acrobatics. Its trainer circled the cord and started to lament while raising his hands in prayer as the cat preened its whiskers whimsically. According to İntizami, this occurred because the cat probably saw a mouse and therefore ended its performance.[40]

The fact that İntizami, who attended the festival himself, observed the secretive sign between the cat and its trainer suggests that it was not intended to be entirely covert in the first place but recognisable by the audience. As Paul Bouissac put forward with respect to contemporary circuses, a set of training codes exists, which are meant to ensure effective communication between the trainer and the performing animal. These codes usually tend to be reduced to a minimum, so it appears that the animal performs autonomously.[41]

Following Bouissac, Michael Peterson pointed out that this kind of code gives the audiences the information that enables them to 'read' what he defines as the 'animal apparatus'. By understanding this apparatus, spectators are able to identify the ways in which animals are made to perform and signify during a performance. The use of a secretive sign, for example, informs the audience about the training the animal has undergone and thus gives a hint about how these performances are realised. Considering the content produced by the performers (both animals and trainers) allows spectators to grasp the concepts of 'animal' or 'animality' as they are manifested during the show.[42]

Thus, by staging a performance with an acrobatic cat that used a training stick and danced gracefully on a cord, the trainer wished to present a different kind of cat that followed a non-animal narrative in which it acted like a human acrobat. Moreover, presenting a cat as an acrobat was meant to introduce its biological superiority over humans: as noted by Bouissac, this is because through varied historical periods, the ability to express acrobatic prowess is considered by spectators to be different from the accepted norm. He adds that such biological superiority is limited to the duration of the performance; when the show ends, spectators will see the acrobats as regular humans outside the context of the circus.[43]

This notion coincides with the idea of carnivalesque enactment. It also relates to the way in which the performance of the cat ended, since it too ceased its acrobatic acts and returned to its usual animal nature. Moreover, in order to reinforce this occurrence for his readers, İntizami suggests that the cat might have seen a mouse – an incident that eventually obliged it to behave in accordance with its own feline nature. This kind of statement allows İntizami to reinforce a firm border between humans and animals outside the framework of performance in order to reassert social norms and boundaries once again.

At certain performances during the Ottoman festival of 1582, the trainers even explicitly discussed the animality of their performing animals as part of the show: in one performance, two trainers entered the field with two bears and began to argue about the nature and abilities of this animal.[44] One of them said that his bear was brave, intelligent and helpful. The other trainer claimed that, although sometimes bears might behave like human beings, when they get angry they roar and thus are always unreliable. Afterwards, the trainers decided to test the bears' nature in front of the audience: İntizami recalls how these animals 'raised their paws, and commenced a fight like two wrestlers whilst trying to trip one another'.[45] Additionally, they 'howled like mountain bears while looking at each other with rage'.[46]

Through their staged discussions regarding bears' nature, the trainers used what Bouissac identifies as a 'code of act', which permits communication between the performer and the audience.[47] Until this point, the trainers made the bears act according to their own animal nature. As a result, they were taught to exhibit a sensation of fear by making frightening noises and fighting with each other. This staged aggression was also part of a well-planned show in which a variety of theatricalised feelings were introduced. This notion is used also during contemporary animal shows. As Peta Tait pointed out, the 'circus took full advantage of how animals are anthropomorphised and, more specifically, of the process by which humans anthropomorphise them with and through their emotions.'[48]

However, in accordance with the carnival atmosphere of the Ottoman festival, spectators eventually were meant to view the bears' performance as a joyous one. As noted by İntizami, the trainers indeed managed to provide this kind of emotional atmosphere. At some point, the bears stopped fighting and started to play jovially with each other: they turned somersaults, went delicately through hoops, and pretended to be young children playing with a wooden rocking horse. Additionally, they acted as if they were shepherds wandering around with their crooks, as well as gardeners strolling through gardens and meadows.[49] Such acts befit Bakhtin's notion of the alternative world of carnival in which the bears behaved like humans. Moreover, while describing the performance, İntizami mentioned that during their act the bears did all these things because 'their animal [nature] got mixed with that of a human being'.[50] By making such a statement, the author once again was suggesting that this kind of mix was appropriate only during performance, that is, during the alternative world of the carnival.

In short, analysing İntizami's descriptions regarding face-to-face encounters with performing animals enables us to observe how they were conceived by the audience both in the real world and during carnival. Keeping this in mind, the special mode of carnivalesque enactment allows us to read *The Imperial Book of Festival* in a new light. Episodes of animal performance bring a range of contrasting emotions to the fore. On one hand, when acting according to their own nature, some animals were meant at first to frighten the audience. On the other hand, when acting like humans these animals generated the emotional sensation of joy, which on several occasions led to laughter among the audience. At some points, this emotional sensation was built upon Ottoman characteristics of humour. This occurred when familiar episodes from daily Ottoman-Muslim life were illustrated: for example, dancing in Arab and Persian styles, or performing as shepherds, a well-known ritual in Ottoman plays.[51]

Since these humorous episodes appear in many other performances in *The Imperial Book of Festival*, it was suggested by Terzioğlu that 'while the Ottoman cultural world was by no means an undifferentiated whole, elite and commoner partook in much the same way in the domain of laughter'.[52] But Terzioğlu's 'elite and commoner' belong to different social (and emotional) communities, and *The Imperial Book of Festival* was created for the benefit only of the elite. This does not mean, of course, that these different groups were unable to understand each other or did not share common features, only that this particular source was meant neither to entertain nor to educate commoners (and their varied emotional communities), but was intended for elite readership.

Therefore, we must also pay attention to the fact that İntizami stressed emotional scenarios associated with the performing animals. These scenarios are related to stories and anecdotes of animals taken from Islamic-Ottoman folklore and myths that were well known among İntizami's social group. Such stories give us the opportunity to grasp the way in which İntizami helped his emotional community to experience the atmosphere of the shows after the event. At certain points, these recounted experiences were slightly different from those ordinary spectators may have felt during the festival due to the fact that they were not part of the animal apparatus that the trainers created during the actual show, but rather İntizami's own additions.

Two important examples are found in the descriptions of the performances of a playful dog and of snake charmers. According to İntizami, a small juggler dog that was 'like Kıtmir, the brave and the faithful keeper of the antechamber' entered the field.[53] He played many games, jumped through many hoops, somersaulted in the air and danced. During its performance, the dog sometimes behaved like a real dog: among other things, it drooled, raced after a jaw bone and put it in its mouth, bared its teeth and dropped its tongue loosely. Yet, at a sign from his patron, the dog continued its acrobatic games. At the end of its performance, this dog went back to its own nature: it opened its mouth to pause for breath, bared its teeth and yet again dropped its tongue loosely.[54]

As we have already learned from the performance repertoires of the cat and bears, the dog's actions were also meant to generate laughter. Thus, for spectators of the festival, this performance produced a joyful atmosphere owing to the non-human narrative that the dog's trainer presented as part of the carnival atmosphere. However, İntizami referred to the performing dog as 'Kıtmir'. This name echoes Sura 18 (The Cave) verses 9–26 from the Koran, in which a dog, named 'Rakim' (translated Kıtmir in Ottoman Turkish) saves a group of Christians from persecution. These verses are interpreted as a Koranic adaptation of the Christian legend of the Seven Sleepers of Ephesus. According to this version, those young men hid in a cave outside the city of Ephesus and fell into a deep sleep for many years while the loyal dog was watching the entrance.[55]

To Muslim believers, the story of the seven sleepers and Kıtmir symbolises one of the signs or miracles whose purpose was to induce wonder at God's creation. According to Islamic tradition, as a reward for his good deeds for mankind, Kıtmir was one of the few animals who gained entry to Paradise. In addition, this dog is known for his human capabilities: while some believe that God gave him the gift of speech, others consider him to be the reincarnation of a human being.[56]

Thus, by naming the performing dog 'Kıtmir', İntizami provides for his readership an emotional scenario of wonder about this animal. During the early modern period, the Ottoman conception of wonder corresponded with the European cognitivist view, according to which this emotion is evoked when the viewer sees something inexplicable.[57] Additionally, as Philip Fisher pointed out, the recognition of wonder occurs when an aesthetic experience elicits thought or religious intelligibility.[58] This kind of thought is exactly what İntizami encouraged while describing the dog's performance for his readers by invoking the renowned Kıtmir.

Another scenario of wonder was used by İntizami while describing one of the most terrifying acts of the festival – the performance of the snake charmers.[59] At this show, skilled charmers interacted with various reptiles in different ways and aroused different emotional attitudes towards them. The author recalls that during this act one of the charmers climbed into a large barrel stuffed with many snakes. At this sight, spectators lost all hope and poured out their woes piteously by saying: 'alas, how tough and strong is the sight of a human who [looks like] an animal'.[60] In order to demonstrate the condition of this charmer, İntizami explains to his readers that this man's face looked like 'The Queen of Serpents' (*Şah-i Maran*). This is another well-known Persian legend that enjoyed great popularity among the Ottomans during the sixteenth century. The Queen of Serpents was a mythological creature with the head of a woman and the body of a serpent who sacrificed her life for the benefit of a human she loved named Cemsab. She allowed him to boil her meat in order to produce a medicine to cure Sultan Keyhusrev when he became ill.[61]

The charmers did not present the performing man as The Queen of Serpents as part of their show. Rather, İntizami mentioned this mythological figure in order to make his readers reflect on the wonder it entailed. Such creatures were also perceived to be part of the natural world created by God – a belief that was highly emphasised by the Ottomans during the sixteenth century, especially in cosmographic manuscripts that deal with the well-known Islamic genre of 'Wonders of Creation' (*Acâibü'l mahlûkât*).[62]

It is important to note, however, that wonder towards animals is by no means an exclusively elite phenomenon. In fact, a similar emotional scenario regarding snakes was introduced to the festival's other spectators as well. In front of their audience, one of the charmers claimed that their profession was serious and not just trickery.[63] In order to prove it, the charmers acted as if they were making an antidote called 'Tiryak' (theriac) out of the snakes' venom.[64] The use of such an antidote was a shared practice that was not unique to the Ottomans or to the Muslim world. However, during the performance in the festival, this practice received a special Islamic meaning: one of the charmers proclaimed out loud that his antidote was called 'Muhammad's Tiryak' because it was made during the days of the Prophet.[65] Thus, by associating the antidote with Muhammad, the snake charmers intended to elicit wonder among spectators as well.

Following the 'world upside-down' principle of carnivals, both the Queen of Serpents and the Tiryak served as cultural generators of wonder, which were meant to introduce snakes differently from their renowned frightening nature. The use of these two different generators allows us to inspect how the emotional scenarios of wonder are tailored to the needs of varied emotional communities. The snake charmers chose to generate wonder through a physical Tiryak made by humans as a reminder of snakes' advantages to mankind.

Reading carefully through this performance's description reveals that the introduction of snakes during this show probably went too far from the desirable atmosphere of carnival. Thus, in order to reinforce this atmosphere once again, the charmers made use of the concept of 'Muhammad's Tiryak', which introduces these reptiles to the audience as wonderful creatures.

On the other hand, İntizami chose to stress wonder by mentioning a mythological creature that was meant to induce wonder at God's creation. As Yehoshua Frenkel argued, this vision of fauna was rooted in the traditional Islamic belief system, in which anecdotes regarding mysterious creatures were not merely meant to amuse but also 'to fortify the social order that the ruling echelons desired'.[66] Therefore, the use of the figure of The Queen of Serpents befits also the political agendas that İntizami emphasised during his work on *The Imperial Book of Festival*, as Sultan Murad's legitimacy to rule was also depicted by this emotional community as God's will.

Conclusion

This chapter proposes a novel approach to reading *The Imperial Book of Festival* by demonstrating the importance of emotions. I have chosen the carnival atmosphere of the Ottoman festival of 1582 to exemplify how feelings towards various performing animals were socially constructed by people rather than by nature. This construction took place on two different levels: firstly, within the actual performances among the acts that the trainers introduced and, secondly, within *The Imperial Book of Festival* as the author İntizami used his own animal repertoires to transmit and elaborate the emotional atmospheres of these shows to his elite readership.

Following a non-animal narrative, animals were presented during performances within an anthropomorphic frame that made them appear and behave differently from their own nature. By staging animals in this manner, the feelings that were elicited towards them were also different: within several examples introduced in this chapter, emotional responses of joy replaced a sensation of fear due to the fact that animals were presented within a state of carnival, of whose limits spectators were consciously aware.

While this applies for almost every animal performance, what is interesting with the Ottoman carnival is how the alternative world that the trainers presented was strictly bound by the time and space in which the performance occurred. During the sixteenth century, the Ottomans highly prized a sense of order and social hierarchy with the sultan at the apex.[67] Therefore, even if the borders between human and animal were consciously blurred within this alternative world, every performance ended in a way that reinforced these borders. This was usually achieved when performers honoured the sultan with prayers and reverences at the end of their act.

As *The Imperial Book of Festival* was designed for elite readership, its emphasis on social hierarchy was stressed in particular. İntizami took advantage of the performing animals in order to depict Sultan Murad as a sovereign who could control nature. Through his descriptions, special emotional scenarios towards animals were described. These scenarios bestowed upon the performances a unique atmosphere that was not included within the animal apparatus that the trainers had constructed. By mentioning renowned anecdotes of marvellous animals, İntizami provided emotional scenarios that his social group highly associated with animals.

One of these emotions was wonder. Contemporary scholars may dismiss the significance that was given to wonderful creatures such as Kıtmir and The Queen of Serpents by arguing that these are merely literary fictions. However, one should remember that modern emotional communities are different from that formed by the Ottoman elite during the period under discussion. For the Ottomans, as well as other medieval and early modern Islamic societies, the existence of all kinds of creatures was meant to convey wonder at God's handiwork. As Syrinx von Hees described, at the time when manuscripts of the Islamic genre of 'Wonders of Creation' were being produced and circulated, all kinds of wonders 'were accepted as transmitted lore and therefore considered part of scientific knowledge'.[68]

During the seventeenth and eighteenth centuries, the Ottoman elite created a different emotional community, which was much more secular than that of the sixteenth century.[69] This new social group's perception regarding wonder at God's creations had weakened. As pointed out by Gottfried Hagen, this change occurred due to Ottoman contacts with Europe, which in turn caused the emergence of another strain of thought.[70] Accordingly, modernisation and the establishment of zoos changed interactions with the animal kingdom permanently and thus the way in which emotions towards animals were experienced and understood by the Ottomans. By this period, another story about Ottoman-animal relations had begun, one that deserves further attention.

Notes

1   The earliest and most complete available manuscript is preserved at the Topkapı Palace Library in Istanbul: Topkapı Palace Museum Library (hereafter TPML), Hazine 1344. See Fehmi Edhem Karatay, *Topkapı Sarayı Müzesi Kütüphanesi Türkçe Yazmalar Kataloğu*, 2 vols (Istanbul: Topkapı Sarayı Müzesi, 1961), i. 232. In this chapter I have consulted the Topkapı's version as it appears in Mehmet Arslan, *Osmanlı saray düğünleri ve şenlikleri*, 3 vols (Istanbul: Sarayburnu, 2008), ii: *İntizami Surnamesi*. Other copies exist: Süleymaniye Library's version (hereafter SL), Hekimoğlu 642; Leiden University Library (Or. 309) and in the Österreichischen Nationalbibliothek in Vienna (HO 70). See Derin Terzioğlu, 'The Imperial Circumcision Festival of 1582: An Interpretation', *Muqarnas*, 12 (1995), 97 n. 4.

2   For a detailed list of royal celebrations from 1298 until 1899, see Arslan, *Osmanlı*, i: *Manzum Surnameler*, 28–30.

3   For a detailed schedule of the festival day by day, see Metin And, *40 Gün 40 Gece-Osmanlı Düğünleri, Şenlikleri, Geçit Alayları* (Istanbul: Toprakbank, 2000), 61–62. See also Nurhan Atasoy, *Surname-i Hümayun: an imperial celebration* (Istanbul: Koçbank Yayınları, 1997), 22–63; Arslan, *Osmanlı*, ii. 77–100.

4 Mikhail Bakhtin, *Rabelais and His World* (Bloomington: Indiana University Press, 1984).

5 James J. Gross and Lisa Feldman Barrett, 'Emotion Generation and Emotion Regulation: One or Two Depends on Your Point of View', *Emotion Review*, 3 (2011), 11.

6 Both kinds of models are the two ends of a continuum that James J. Gross and Lisa Feldman Barrett suggested in order to illustrate a diversity of perspectives on emotions. This continuum offers other points of view regarding emotion generation and regulation, including appraisal, as well as psychological construction models. For further details see Gross and Barrett, 'Emotion Generation', 8–16.

7 Ibid., 10. See also Sara Ahmed's discussion on the nature of emotions according to 'inside-out models' and 'outside-in models', in *The Cultural Politics of Emotion* (2nd ed., New York: Routledge, 2015), 8–12.

8 Barbara H. Rosenwein, *Generations of Feeling* (Cambridge: Cambridge University Press, 2016), 1; id., 'Worrying about Emotions in History', *American Historical Review*, 107 (2002), 824–26. See also Dror Wahrman's study on the transformations of human behaviour and cultural history studies, in his 'Change and the Corporeal in Seventeenth- and Eighteenth-Century Gender History: Or, Can Cultural History Be Rigorous?', *Gender & History*, 20 (2008), 584–602.

9 Barbara H. Rosenwein, *Emotional Communities in the Early Middle Ages* (London: Cornell University Press, 2007), 1–31.

10 Barbara H. Rosenwein, 'Theories of Change in the History of Emotions', in Jonas Liliequist (ed.), *History of Emotions, 1200–1800* (London: Pickering and Chatto, 2012), 12.

11 Rosenwein, *Generations of Feeling*, 3.

12 Walter G. Andrews, 'Ottoman Love: Preface to a Theory of Emotional Ecology', in Jonas Liliequist (ed.), *History of Emotions* (London: Pickering and Chatto, 2012), 23–24.

13 Ibid., 23.

14 Arslan, *Osmanlı*, ii. 210.

15 'Bu bir şīr-i mehīb ki bir ḥamlesinden burc-ı ḥamele zelzele ve bir şadmesinden şīr-i felege velvele düşer...yerinde ki Fīl-i Mahmūdī eline girse...karşu pençesi ile yırtar...' Ibid., ii. 210.

16 'Fe-emmā ne deñlü ḥayvân ise mehâbet-i pâdişâhīden kendüyi teslīm menziline tenzīl idüp...' Ibid., ii. 210.

17 Paul A. Trout, *Deadly Powers: Animal Predators and the Mythic Imagination* (New York: Prometheus Books, 2011), 67.

18 Ibid., 67–68.

19 Rod Preece, *Awe for the Tiger, Love for the Lamb: A Chronicle of Sensibility to Animals* (New York: Routledge, 2002), 5.

20 Johannes Leunclavius (1541–1594) was a German historian and orientalist who took part in the festival and described the battle using this symbolic interpretation. Atasoy, *Surname-i Hümayun*, 107.

21 'şīr-bânlar zebân-ı melâmeti dıraz eylediler ve şan'atlarınuñ kem ü kâstların barmaġa ṭoladır' Arslan, *Osmanlı*, ii. 210.

22 *Encyclopaedia of Islam*, ed. P. J. Bearman et al., 12 vols (2nd ed., Leiden: Brill, 1960–2005) [hereafter *EI*], s.v. 'Malāmatiyya'.

23 Dacher Keltner and Jonathan Haidt, 'Approaching Awe, A Moral, Spiritual, and Aesthetic Emotion', *Cognition and Emotion*, 17 (2003), 299.

24 Serpil Bağcı, 'Visualizing Power: Portrayals of the Sultan in Illustrated Histories of the Ottoman Dynasty', *Islamic Art*, 6 (2009), 118.

25 Suraiya Faroqhi, *Subjects of the Sultan: Culture and Daily Life in the Ottoman Empire* (London: I. B. Tauris, 1995), 164–65. For further information regarding the literary genre of books of festivals see Arslan, *Osmanlı*, i. 23–134; Suraiya Faroqhi and

Arzu Öztürkmen, 'Research on Festivals and Performances', in Suraiya Faroqhi and Arzu Öztürkmen (eds), *Celebration, Entertainment and Theater in the Ottoman World* (London: Seagull Books, 2014), 24–70.

26   About Murat's book purchases to the Topkapı Library, see: Atasoy, *Surname-i Hümayun*, 12.

27   Ibid., 14–15.

28   Emine Fetvacı, *Picturing History at the Ottoman Court* (Indianapolis: Indiana University Press, 2013), 176–77. See also Baki Tezcan, *The Second Ottoman Empire* (Cambridge: Cambridge University Press, 2010), 101–8.

29   Fetvacı, *Picturing History*, 176–77.

30   Ibid., 181–83.

31   Alan Mikhail, *The Animal in Ottoman Egypt* (New York: Oxford University Press, 2014), 109–36.

32   *Türkiye Diyanet Vakfı İslam Ansiklopedisi* (Istanbul: Türkiye Diyanet Vakfı, Islâm Ansiklopedisi Genel Müdürlüğü, 1988-) [hereafter *TDVİA*], s.v. 'Arslanhane'.

33   Bakhtin, *Rabelais*, 71.

34   Ibid., 5–6.

35   Fariba Zarinebaf, 'Asserting Military Power in a World Turned Upside Down', in Suraiya Faroqhi and Arzu Öztürkmen (eds), *Celebration, Entertainment and Theater* (Chicago: University of Chicago Press, 2014), 173–85.

36   Terzioğlu, 'Imperial Circumcision Festival', 96.

37   Shulamith Lev-Aladgem, 'Carnivalesque Enactment at the Children's Medical Centre of Rabin Hospital', *Research in Drama Education*, 5 (2000), 166.

38   This festival took place at the archery area of *Okmeydanı* in Istanbul for the celebrations of Sultan Ahmed III (r.1703–1730) four sons' circumcisions. About this Book of Festival, see TPML, Ahmet III 3593 and Ahmet III 3594; Karatay, *Topkapı Sarayı Müzesi*, i. 280–81; Esin Atil, *Levni and the Surname: The Story of an Eighteenth-Century Ottoman Festival* (Istanbul: Koçbank, 1999). For a detailed list of other copies of this *Surname*, see: Arslan, *Osmanlı*, iii. 12. For this chapter I have consulted the copy of İstanbul University Library (hereafter İUL), TY. 6099, as it appears in Arslan Arslan, *Osmanlı*, iii. 207.

39   'erbâb-ı temâşâ meslubü'ş-şu'r olacaḳ mertebe kesb-i şeṭâret idüp gülüşdiler'. Arslan, *Osmanlı*, iii. 207.

40   Arslan, *Osmanlı*, ii. 327. See also Atasoy, *Surname-i Hümayun*, 111.

41   Paul Bouissac, *Circus and Culture* (Bloomington: Indiana University Press, 1976), 52–57.

42   Michael Peterson, 'The Animal Apparatus: From a Theory of Animal Acting to an Ethics of Animal Acts', *The Drama Review*, 51 (2007), 34–35.

43   Bouissac, *Circus and Culture*, 46–50.

44   Arslan, *Osmanlı*, ii. 428–29. See also Atasoy, *Surname-i Hümayun*, 108.

45   'Pes ol iki ḥırs-ı dilīr…birbirine pençe şalup el virişdiler ve küştī-gīr pehlevânlar gibi iki ayaġ üzere ṭurup birbirine şarmadan pâ-bend geçmege ḥayli dürüşdiler'. Arslan, *Osmanlı*, ii. 428–29.

46   'ṭaġ ayular gibi ol dü-pâ ḥırslar birbirine ġażabla baḳarken vâfir uluşdılar'. Ibid., 429.

47   Bouissac, *Circus and Culture*, 52.

48   Peta Tait, *Wild and Dangerous Performances: Animals, Emotions, Circus* (Basingstoke: Palgrave Macmillan, 2012), 1.

49   Arslan, *Osmanlı*, ii. 429.

50   'cân-verler iken âdame ḳarışmaġla'. Ibid., 429.

51   Metin And, *Drama at the Crossroads: Turkish Performing Arts Link Past and Present, East and West* (Istanbul: Isis Press, 1991), 45–46.

52   Terzioğlu, 'Imperial Circumcision Festival', 97.

53 'İşbu ḳıṭmīr, a'nī seg-i dilīr, pehlevân-ı meydân-ı mübârezet ve pâs-bân-ı eyvân-ı muḥâzeret'. Arslan, *Osmanlı*, ii. 438. See also Atasoy, *Surname-i Hümayun*, 109.

54 Ibid., 439.

55 Metin And, *Minyatürlerle Osmanlı- İslam Mitologyası* (Istanbul: YKY, 2012), 232–37; *EI*, s.v. 'Aṣḥāb al-Kahf'; *TDVİA*, s.v. 'Ashab-ı Kehf'.

56 *EI*, s.v.'Kalb'.

57 Caroline Walker Bynum, 'Wonder', *American Historical Review*, 102 (1997), 1–26.

58 Philip Fisher, *Wonder, the Rainbow, and the Aesthetics of Rare Experiences* (Cambridge: Harvard University Press, 1998), 37–38.

59 Arslan, *Osmanlı*, ii. 207; SL, Hekimoğlu 642, fos. 69a–70a.; Atasoy, *Surname-i Hümayun*, 54–55.

60 'âyâ bu ne ḥâletdür, ne it cânlu ṣūret-i insânda ḥayvânlar olur imiş'. Arslan, *Osmanlı*, ii. 207.

61 And, *Minyatürlerle Osmanlı- İslam Mitologyası*, 63–64; *TDVİA*, s.v. 'Câmasbnâme'.

62 This genre enjoyed great popularity throughout the Islamic world, especially after the thirteenth century when the Arab cosmographer and geographer al-Qazwini (1203–1283) composed his work known as *"Adjā'ib al-makhlūḳāt wa-gharā'ib al-mawdjūdāt'* ('Prodigies of things created and miraculous aspects of things existing'). *EI*, s.v. 'al-Ḳazwīnī'; For a detailed list of 'Wonders of Creation' manuscripts produced by the Ottomans as well as by other Islamic societies, see Persis Berlekamp, *Wonder, Image, and Cosmos in Medieval Islam* (New Haven: Yale University Press, 2011), 179–84.

63 Arslan, *Osmanlı*, ii. 207.

64 This kind of antidote was considered to be the most complex of Muslim pharmaceutical forms, because it contained a huge number of ingredients. See Miri Shefer Mossensohn, *Ottoman Medicine: Healing and Medical Institutions 1500–1700* (Albany: SUNY, 2009), 35; Leigh Chipman, *The World of Pharmacy and Pharmacists in Mamlūk Cairo* (Leiden: Brill, 2009), 282.

65 Medieval Muslim culture dedicated great attention to theriac. It was especially famous among the Mamluks who used it as a gift item for European monarchs. See Doris Behrens-Abouseif, *Practising Diplomacy in the Mamluk Sultanate: Gifts and Material Culture in the Medieval Islamic World* (London: I. B. Tauris, 2014), 148–49.

66 Yehoshua Frenkel, 'Animals and Otherness in Mamluk Egypt and Syria', in Francisco de A. García, Mónica A. Walker Vadillo and María V. Chico Picaza (eds), *Animals and Otherness in the Middle Ages* (Oxford: Archaeopress, 2013), 56.

67 Fetvacı, *Picturing History*, 178–79.

68 Syrinx von Hees, 'The Astonishing: A Critique and Re-Reading of 'Aǧā'ib Literature', *Middle Eastern Literatures*, 8 (2005), 105–6.

69 Baki Tezcan, 'The Second Empire: The Transformation of the Ottoman Polity in the Early Modern Era', *Comparative Studies of South Asia, Africa and the Middle East*, 29 (2009), 569–70.

70 Gottfried Hagen, 'Ottoman Understanding of the World in the Seventeenth Century', in Robert Dankoff (ed.), *An Ottoman Mentality* (Leiden: Brill, 2004), 226.

Bibliography

And, Metin, *Drama at the Crossroads: Turkish Performing Arts Link Past and Present, East and West* (Istanbul: Isis Press, 1991).

———, *40 Gün 40 Gece-Osmanlı Düğünleri, Şenlikleri, Geçit Alayları* (Istanbul: Toprakbank, 2000).

———, *Minyatürlerle Osmanlı-İslam Mitologyası* (Istanbul: YKY, 2012).

Andrews, Walter G., 'Ottoman Love: Preface to a Theory of Emotional Ecology', in Jonas Liliequist (ed.), *History of Emotions* (London: Pickering and Chatto, 2012), 21–47.

Arslan, Mehmet, *Osmanlı saray düğünleri ve şenlikleri*, 3 vols (Istanbul: Sarayburnu, 2008).

Atasoy, Nurhan, *Surname-i Hümayun: an imperial celebration* (Istanbul: Koçbank Yayınları, 1997).

Atil, Esin, *Levni and the Surname: The Story of an Eighteenth-Century Ottoman Festival* (Istanbul: Koçbank, 1999).

Bakhtin, Mikhail, *Rabelais and His World* (Bloomington: Indiana University Press, 1984).

Behrens-Abouseif, Doris, *Practising Diplomacy in the Mamluk Sultanate: Gifts and Material Culture in the Medieval Islamic World* (London: I. B. Tauris, 2014).

Berlekamp, Persis, *Wonder, Image, and Cosmos in Medieval Islam* (New Haven: Yale University Press, 2011).

Bouissac, Paul, *Circus and Culture* (Bloomington: Indiana University Press, 1976).

Bynum, Caroline Walker, 'Wonder', *American Historical Review*, 102 (1997), 1–26.

Chipman, Leigh, *The World of Pharmacy and Pharmacists in Mamlūk Cairo* (Leiden: Brill, 2009).

Dankoff, Robert (ed.), *An Ottoman Mentality* (Leiden: Brill, 2004).

Encyclopaedia of Islam, ed. P. J. Bearman et al., 12 vols (2nd ed., Leiden: Brill, 1960–2005).

Faroqhi, Suraiya, *Subjects of the Sultan: Culture and Daily Life in the Ottoman Empire* (London: I. B. Tauris, 1995).

————— and Arzu Öztürkmen (eds), *Celebration, Entertainment and Theater in the Ottoman World* (London: Seagull Books, 2014).

—————, 'Research on Festivals and Performances', in Suraiya Faroqhi and Arzu Öztürkmen (eds), *Celebration, Entertainment and Theater in the Ottoman World* (London: Seagull Books, 2014), 24–70.

Fetvacı, Emine, *Picturing History at the Ottoman Court* (Indianapolis: Indiana University Press, 2013).

Fisher, Philip, *Wonder, the Rainbow, and the Aesthetics of Rare Experiences* (Cambridge: Harvard University Press, 1998).

Frenkel, Yehoshua, 'Animals and Otherness in Mamluk Egypt and Syria', in Francisco de A. García, Mónica A. Walker Vadillo and María V. Chico Picaza (eds), *Animals and Otherness in the Middle Ages* (Oxford: Archaeopress, 2013), 49–61.

García, Francisco de A., Mónica A. Walker Vadillo and María V. Chico Picaza (eds), *Animals and Otherness in the Middle Ages* (Oxford: Archaeopress, 2013).

Gross, James J. and Lisa Feldman Barrett, 'Emotion Generation and Emotion Regulation: One or Two Depends on Your Point of View', *Emotion Review*, 3 (2011), 8–16.

Hagen, Gottfried, 'Ottoman Understanding of the World in the Seventeenth Century', in Robert Dankoff (ed.), *Ottoman Mentality* (Leiden: Brill, 2004), 215–56.

Karatay, Fehmi Edhem, *Topkapı Sarayı Müzesi Kütüphanesi Türkçe Yazmalar Kataloğu*, 2 vols (Istanbul: Topkapı Sarayı Müzesi, 1961).

Keltner, Dacher and Jonathan Haidt, 'Approaching Awe, a Moral, Spiritual, and Aesthetic Emotion', *Cognition and Emotion*, 17 (2003), 297–314.

Kirillina, Svetlana, 'Representations of Animal World of the Ottoman Empire in Russian Christian Pilgrims' Accounts', in Suraiya Faroqhi (ed.), *Animals and People in the Ottoman Empire* (Istanbul: Eren, 2010), 75–97.

Lev-Aladgem, Shulamith, 'Carnivalesque Enactment at the Children's Medical Centre of Rabin Hospital', *Research in Drama Education*, 5 (2000), 163–74.

Liliequist, Jonas (ed.), *History of Emotions, 1200–1800* (London: Pickering and Chatto, 2012).

MacLean, Gerald, *Looking East* (New York: Palgrave Macmillan, 2007).

Mikhail, Alan, *The Animal in Ottoman Egypt* (New York: Oxford University Press, 2014).

Peterson, Michael, 'The Animal Apparatus: From a Theory of Animal Acting to an Ethics of Animal Acts', *The Drama Review*, 51 (2007), 33–48.

Rosenwein, Barbara H., 'Worrying about Emotions in History', *American Historical Review*, 107 (2002), 821–45.

———, *Emotional Communities in the Early Middle Ages* (London: Cornell University Press, 2007).

———, 'Theories of Change in the History of Emotions', in Jonas Liliequist (ed.), *History of Emotions, 1200–1800* (London: Pickering and Chatto, 2012), 7–20.

———, *Generations of Feeling* (Cambridge: Cambridge University Press, 2016).

Shefer Mossensohn, Miri, *Ottoman Medicine: Healing and Medical Institutions 1500–1700* (Albany: SUNY, 2009).

Tait, Peta, *Wild and Dangerous Performances: Animals, Emotions, Circus* (Basingstoke: Palgrave Macmillan, 2012).

Terzioğlu, Derin, 'The Imperial Circumcision Festival of 1582: An Interpretation', *Muqarnas*, 12 (1995), 84–100.

Tezcan, Baki, 'The Second Empire: The Transformation of the Ottoman Polity in the Early Modern Era', *Comparative Studies of South Asia, Africa and the Middle East*, 29 (2009), 556–72.

———, *The Second Ottoman Empire* (Cambridge: Cambridge University Press, 2010).

Trout, Paul A., *Deadly Powers: Animal Predators and the Mythic Imagination* (New York: Prometheus Books, 2011).

Türkiye Diyanet Vakfı İslam Ansiklopedisi (Istanbul: Türkiye Diyanet Vakfı, Islâm Ansiklopedisi Genel Müdürlüğü, 1988-).

von Hees, Syrinx, 'The Astonishing: A Critique and Re-Reading of 'Aǧā'ib Literature', *Middle Eastern Literatures*, 8 (2005), 101–120.

Wahrman, Dror, 'Change and the Corporeal in Seventeenth- and Eighteenth-Century Gender History: Or, Can Cultural History Be Rigorous?', *Gender & History*, 20 (2008), 584–602.

Zarinebaf, Fariba, 'Asserting Military Power in a World Turned Upside Down', in Suraiya Faroqhi and Arzu Öztürkmen (eds), *Celebration, Entertainment and Theater* (Chicago: University of Chicago Press, 2014), 173–85.

2 Sleeve cat and lap dog

Affection, aesthetics and proximity to companion animals in Renaissance Mantua

Sarah Cockram[1]

In 1512, Isabella d'Este, marchesa of Mantua in northern Italy, was given a tiny exotic pet cat from India. This rare red cat was known as the *animalino* (little animal),[2] and he was small enough to be carried in Isabella's sleeve. Portable, exceptional and adorable, the *animalino* was an enviable addition to Isabella's famous collections of the finest possessions. The *animalino* was a luxury but also a sentient creature with lived experience in often-changing surroundings, a creature habituated to human contact and touch, and one who provoked a range of feelings in the people around him, from envy to awe to joy and affection.

In this chapter, I examine the case of the *animalino*, and other companion cats and dogs at the Mantuan court. Adding to a flourishing body of literature on the history of the human–pet relationship,[3] I argue that living with companion animals was not only an expression of status and identity, but also a matter of aesthetics, sensory delight, and human feelings, and I explore the possibility of understanding animal experience and feelings as equally important considerations. In line with the aims of this volume as a whole, this chapter prioritises interspecies interaction and 'entwinement'.[4] While acknowledging a reliance on human sources, it is committed to taking the animal perspective seriously and to raising issues of both human and animal wellbeing.

Within the framework of interdisciplinary historical animal studies, this chapter flags up opportunities and challenges of analysing interspecies interactions in the past through different methodological approaches. The most obvious of these for the historian (and certainly revealing here) is the use of evidence found in correspondence and other documents. Another way to investigate affective and haptic bonds between people and companion animals in late medieval and early modern Europe, in this case in Renaissance Italy, is to zoom in on the hand in the fur in visual representations of the human–animal relationship. This can open up analysis of the aesthetic, tactile, and affective importance of proximity and the stroking of an animal, as well as that animal's response to human touch. This chapter also highlights where archaeological evidence can be valuable and engages with modern research from social, medical, and animal sciences. As Erica Fudge's recent article shows in using the work of the American animal scientist Temple Grandin to re-consider the sensory experience of seventeenth-century cows in England, when responsibly done and culturally situated, creative application of ideas from across disciplines to our historical sources can offer valuable new perspectives on 'human-animal shared worlds', present and past.[5]

For instance, with requisite caveats of historical/cultural context, a modern study by the sociologist Krzysztof Konecki presents a useful tool for interpreting interspecies contact. Konecki enumerates four kinds of positive communicative touching of a pet.[6] These include 'ritual touching', for instance in patting an animal in greeting or farewell; 'controlling touching', as in Titian's well-known portrait of Charles V with his hand on the collar of a large dog (1533, Prado, Madrid); and 'entertaining touching': in other words, playing. The final kind of communicative touching of a pet is 'touching characterised by positive affection'. This is found frequently throughout the examples in this chapter, although the four types of touching are, of course, often interlinked and one touch might fit in more than one category at a given time.

To argue for the centrality of affection, proximity, and sensory engagement in the human-companion animal relationship, I begin by discussing the ubiquity of high-status companion animals at Renaissance courts, in particular the court of Mantua studied here, before introducing the *animalino* of this chapter's title and other individual dogs and cats belonging to several members of the ruling Gonzaga family. This chapter demonstrates the significance of aesthetics and delight in the relationship of owners to these courtly creatures, as well as these animals' importance to the sensescape of the court environment,[7] and discusses factors of animal agency, behaviour, and personality. This is followed by a call to consider implications for the health and wellbeing of owner and animal in this relationship, including the impact of bereavement. The chapter closes by returning to the haptic relationship and mutual touch. What did it mean to live with, and love, lapdogs and sleeve cats in Renaissance Mantua? Were these animals reduced to being, to quote Donna Haraway, 'affectional slaves'?[8] Or might pet-keeping have brought significant bonds that were good for both human and animal?

Power and pets in Renaissance Mantua

Isabella d'Este (1474–1539), owner of the *animalino*, left her home state of Ferrara to join her marital family in 1490, as the fifteen-year-old bride of Francesco II Gonzaga (1466–1519), fourth marquis of Mantua. Working together for almost three decades until Francesco's death, the couple navigated Renaissance Italy's tumultuous political landscape,[9] and were able to pass on their state to their eldest son Federico (1500–40), who would become Mantua's first duke in 1530. Isabella and Francesco's strategy of co-rule included the construction of a court that projected power through magnificence.

Animals were a key feature of this. According to Baldesar Castiglione, a great prince should

> hold magnificent banquets, festivals, games and public shows, and keep a great many fine horses for use in peace or war, as well as falcons, hounds and all the other things that pertain to the pleasures of great lords and their subjects: after the manner of [Francesco] Gonzaga ... in our own day, who in this regard seems more like King of Italy than the ruler of a city.[10]

The breeding and collecting of impressive court animals magnified the image of Gonzaga splendour and allowed the family to build connections through gift-giving, often providing access and lines of communication to the most powerful international rulers such as the Ottoman sultan, the pope, the emperor or the kings of France and England.[11] In addition to the famous Gonzaga horses, exotic animals, raptors, and hunting dogs, companion species had a central role in this gift economy. The court of Mantua was associated with such creatures, the Gonzaga having several dog-related devices and being renowned for the quality of companion animals at their court, including cats.[12] In addition to the importance of companion animals in displays of power, including through expensive accessories,[13] sentiment also frequently connected rulers to their favourites. The research of Rodolfo Signorini has given the name Rubino to the large red hunting dog who is settled under the chair of Lodovico II Gonzaga (grandfather of Francesco II) in Mantegna's *Camera degli Sposi*, also known as the *Camera Picta* (Figure 2.1).

Letters make clear the affective bond between Lodovico (1412–78) and his dog. They document the prince's desperation to find Rubino when in 1462 he ran away in the countryside looking for his master (only to turn up soaking wet that evening back at the palace); Rubino's anxiety, pacing room to room in search of Lodovico; the reunion of lost dog and master with 'lots of caresses'; as well as Lodovico's concern during Rubino's final illness and distress at his death in 1467.[14]

Amiable though he looks in Mantegna's portrait, Rubino was apparently unpredictable, needing to be chained when not with Lodovico. There is a very clear correlation, of course, between rulers and powerful animals – as seen in Titian's portrait of Charles V mentioned above. In this association is displayed the ability of the prince to tame fearsome creatures.[15] According to Pierre Belon, Francis I of France was said to have had a lion or other ferocious creature that slept on his bed as his subjects would have a dog.[16] Such extraordinary companions could confirm the power and nature of the ruler (or modern-day celebrity or president) through the respect, loyalty, and bond accorded this special human by the animal. They indicated wealth, broad horizons, and control over subjects and dominions. Fierce creatures could be given as high-profile gifts and add beauty and wonder to a court.[17] Status is a factor but often so too were companionship and enjoyment: Francis's lion was held dear and provided company and pleasure.

Figure 2.1 Rubino, the favourite dog of Marchese Ludovico Gonzaga III [also known as Lodovico II] of Mantua and his family, from the *Camera degli Sposi* or *Camera Picta*, 1465–74 (fresco) (detail of 78461), Mantegna, Andrea (1431–1506) / Palazzo Ducale, Mantua, Italy / Bridgeman Images.

To quote James Serpell: 'Throughout history the world's ruling classes have almost invariably demonstrated a powerful affinity for pets'.[18] This attachment is often expressed in proximity (as with the lion on the king's bed or Rubino under Lodovico's chair), profound understanding between human and animal, and a sense that other people could not be trusted in the same way.[19] Contact with such animals might encompass several of Konecki's types of touch, from greeting to play, control, or affection. There is thus often more to the relationship between ruler and animal than simply the manifestation of magnificence, as for instance when Gian Galeazzo Sforza, young duke of Milan, asked for his favourite horses and hounds to be brought to his deathbed.[20]

Superb companion animals need not be large or fierce to show power. As with the dogs and cats in this chapter, novelty, fine-breeding, aesthetic delight, charming behaviour, and a place as the recipient or giver of displays of affection could contribute to the court animal's value (symbolic, monetary, and emotional). Affective bonds are revealed by examining the relationship between rulers and small companion animals, such as lapdogs. Although lapdogs may lack the obvious utility of hunting dogs, their traditional primary occupation was in providing pleasing companionship – a role not to be dismissed. In addition to showing emotional connections between the Gonzaga and their lapdogs, this chapter aims to add more nuance to the stereotypical view that associates big dogs more or less exclusively with male owners and little dogs with ladies.[21] As well as his large dogs such as Rubino, Lodovico Gonzaga, for example, was fond of smaller dogs, including a toy spaniel named Bellina, and Lodovico's great-grandson Federico II Gonzaga was attached to a succession of favourite lapdogs throughout his life.

Federico's mother, Isabella d'Este, was known for dogs of various kinds, including spaniels and bichons. Now famous as a shrewd stateswoman, a 'Machiavelli in skirts' and, above all, a patroness of music and art,[22] Isabella was also a collector of fine animals. These included Gonzaga horses, civet cats bred for their musk, and parrots. She took a personal interest in dogs, big and small, including overseeing their breeding or acquisition, their training, and their naming.[23] Isabella's most famous dog is Aura, who will be considered further below. Aura accompanied Isabella in her daily life, lived in her private rooms, and, we are told, wagged her tail and affectionately chewed on the marchesa's fingers.[24] But we turn first to Isabella's unusual cat, the *animalino*.

The *animalino*

The cat known in correspondence as the *animalino* was sent to Isabella in 1512 as a prestigious gift from a twelve-year-old Federico. The juvenile Gonzaga heir was then in Rome, nominally a hostage to Pope Julius II but in effect occupying the position of papal favourite. Federico described the kitten he had managed to procure as 'the most beautiful and most delicate little cat that has ever come from Calicut'. The animal had been a gift to the Roman patrician Angelo del Buffalo from the queen of Portugal and Angelo swore to Federico that no such animal had been seen before in Italy. This rare 'small, delicate and marvellous little animal', handily sleeve-sized, was jealously sought by all who saw him.[25] Federico recognised the *animalino* as the ideal gift for his mother, whose reputation for singularity was based on the possession of such marvels and whose attachment to matchless companion animals was well known. The *animalino* was to be sent to Mantua.

Contemporary images show various types of cats, and Isabella certainly acquired felines for different purposes, including cats from Spain and the Levant.[26] The *animalino* is described as 'rosso', and whether he was a rare red

Eastern domestic cat or a tamed wild cat is still unclear. It will be helpful if continued investigation allows a conclusion to be reached about his species, permitting better analysis of this cat's needs, experiences, and health. If further archival and on-site research allows identification of a burial site for the *animalino*, DNA analysis could provide vital evidence.[27] In the absence of such confirmation, several hypotheses can be put forward. One option to consider is that he may have been a young Pallas' cat (*Otocolobus manul*). Also known as a manul, this species is usually about the size of a domestic cat so is perhaps a little large for our *animalino*. On the other hand, the manul has distinctive plush fur (reddish in its erythristic phase) and is noted for its appealing, kittenish features. Another candidate is the elusive Rusty Spotted Cat (*Prionailurus rubiginosus*), one of the smallest cats in the world and thus a good fit for size.[28]

Figure 2.2 Bacchiacca, *Woman with a Cat* (1540s), ART Collection / Alamy Stock Photo.

Whatever the species of the *animalino*, and it may well be that he was an unusual domestic cat of some kind, such as that in Bacchiacca's portrait of a lady with a cat (Figure 2.2), Isabella awaited his arrival with great anticipation. Her words give a clue to her excitement and some of her motivations, she was: 'in incredible desire as it is of the novelty, beauty, and delicateness that you tell us'.[29] The following month, July 1512, the cat not yet dispatched, Isabella gladly accepted her son's offer to send two more cats with the *animalino*, a little black male and a female. The black male died on the journey, when he got caught between the horse who was carrying him and a column to which the horse was secured. Isabella wrote to her son in expectation that 'that beautiful and singular *animalino*' arrive safely with the other. She told her son she would treasure the creature. She said she would look after this cat and care for him personally to remember Federico and for her own great pleasure.[30] So she waited to take proud ownership of this animal for his uniqueness, aesthetic appeal, connection to her son, and her own enjoyment.

Once in her possession after safe arrival on 13 August, Isabella was thrilled with the *animalino*.[31] She carried him at her breast and in her sleeve all that first day. On this point it should be stressed that companion animals were often kept in close proximity to the owner within their clothing. Mary, Queen of Scots had an entourage of lapdogs, dressed in blue velvet in winter, and when she was executed famously had a dog hiding under her skirts.[32] Fashionable wide sleeves combined with the daintiness desirable in certain companion animals meant that wearing a small pet in one's sleeve was eminently possible. The sleeve was an important site of display,[33] and the relationship of the living animal within the sleeve to dead fur that may line the sleeve is of interest,[34] as is the sensory appeal of the warm animal nestled within one's clothing. The act of carrying a tiny cat in one's sleeve – as with having in hand or on one's person any delicate, precious object such as fine glassware or intricate accessories – would also highlight the elegance of one's movement, in a display of that quality of *sprezzatura* described in relation to Castiglione's ideal courtier.[35] This would require poise that was carefully controlled yet seemingly effortless, and to be successful would demand co-operation from the animal partner. The sentience and agency of these particular luxury possessions were thus prime considerations.

Once the *animalino* was out of Isabella's sleeve, she showed him with pride to her husband Francesco, who was attending celebrations at the cathedral in Mantua. Crowds marvelled at the *animalino,* now in the prince's arms. This was a cat to be fussed over, carried, and admired,[36] receiving that type of touch Konecki terms 'positive affection'. In his new home, Isabella ensured that the *animalino* rested in comfort and he had a special little bed. While some cats had freedom to rat and roam in the palace (and other courts provide evidence of cat flaps for this purpose),[37] the *animalino* was presumably kept under close watch and a degree of confinement in Isabella's rooms. He was apparently given food he liked (future research may establish what he ate) and Isabella never ceased to kiss him and praise him with tender words and baby talk. Such motherese and mothering, proximity and touch, would characterise this new relationship, and her secretary described Isabella's love for her new favourite companion animal.

The third little cat sent from Rome, the female, was also feted and lodged in Isabella's room, but 'blessed' was anyone allowed near enough to touch the *animalino*. This was the prerogative of the ruling couple. The *animalino* seems well behaved to stay so peaceably within the sleeve and in new hands in the busy environment of the court – not to mention the cathedral and the piazza in between. He may have been easy natured, or well trained in advance of his arrival, or there may have been more concerning reasons for his calm behaviour.

The *animalino*'s arrival in Mantua came after a series of potentially stressful journeys – India to Portugal to Rome to Mantua via Florence – the last leg, at least, in intense heat and during which a feline travelling companion was killed. In his time before Mantua, he presumably must also have been regularly and gently handled to produce the tractable creature of whom Isabella took receipt. Once installed in her rooms his basic needs and more were met: for food, comfort, and perhaps affection – but how can we know if he liked the stroking, the attention, and the carrying in a sleeve? One could argue that if he did not we might find examples of recalcitrance and resistance in his behaviour. He might attempt escape (even successfully, as did a Syrian cat sourced for Isabella in Venice that leapt from a balcony and made off, never to be dispatched to Mantua after all)[38] or he might wriggle, scratch, or bite. Such a troublesome *animalino* would not likely have got so far as to be presented to Isabella in the first place. The *animalino* may then have been content with his situation – or not, and we will return to his wellbeing as he settled in to his new home later.

Delight, aesthetics, and the senses

What of Isabella? Aside from the status-symbol factor, clearly significant to Isabella in wanting this delicate and exotic animal, what else did she get out of the relationship? One, utilitarian, argument often put forward for the value of companion animals in the past, and a reason one might be kept inside the clothes, was that these could attract parasites away from the body. Juliana Barnes in the 1486 *Boke of Saint Albans* mentions 'smale ladies popis that beere a way the flees'.[39] Beyond the *animalino*-as-flea-catcher or the other argument made for companion animal utility of the time, that they could function as a sort of compress or living hot water bottle (perhaps not desirable in August in Mantua),[40] what other benefits might ownership of a tiny Eastern cat have had for Isabella's wellbeing? There may have been psychological benefits to babying the creature and bonding with him as she did. The *animalino* connected Isabella with present and absent loved ones: with her beloved son Federico and with her husband Francesco, with whom marital relations were strained at that time.[41] These Gonzaga connections through companion animals are also found in ideas of social and family groups that included each other's dogs and cats.[42] For instance, Isabella presented to Federico a pretty spaniel puppy, offspring of her dog, to be named Zephyro in a wind allusion to match Aura.[43] Later Zephyro would sire puppies by Isabella's dog Mamia.[44]

Isabella took joy from the *animalino*'s company, in the same way that she referred to the companion animals of a friend in a letter of 1493. Hearing that Beatrice da Trani was dying, Isabella wrote to a mutual friend asking that when Beatrice pass away, Isabella be sent her black dog and 'that beautiful cat which so delighted her'.[45] Delight could be derived from the personality, spontaneity, and behaviour of the companion animal and from the personal relationship of the owner with the animal as an individual as well as a type,[46] as seen in the examples of the lovable docility of the *animalino*, or Aura's playfulness and friendly bites, or in the endearing single-minded devotion of Rubino – running away to find his master, pining and refusing to behave for others.[47] Companion animal personality raises questions of the inter-relationships between training and agency and between breeding and individuality.[48]

Isabella shows us that delight in the company of Renaissance companion animals might be derived from several sources. As well as from animals' characters, delight came from rarity, beauty, and other sensory pleasure related to proximity, touch, and stroking. The *animalino* was prized for his exceptional pelt of rare colour and appeal, and Isabella otherwise sought cats for softness and fine markings.[49] The relationship linking aesthetics, fashions, and the breeding and selection of animal bodies is one that will benefit from future research into different times and places. There is evidence for Renaissance Italy of the breeding of luxury animals specifically for strokability and of responses to such qualities. For example, the Bolognese naturalist Ulisse Aldrovandi gives advice on how to breed smaller dogs with longer coats.[50] Fur is important as a commodified luxury item removed from animals, but there is also a significance in living fur and the beauty ideal for certain types of companion animals, which featured a variety of sensory dimensions. This ideal includes attributes of appearance, such as smallness, cuteness, and infantile characteristics, lustrousness, and shine,[51] sometimes whiteness or other colours and markings. These characteristics can be reflected in dogs' names, which frequently include a diminutive suffix: Bellina ('little pretty one'), Perlina ('little pearl', a Medici dog), or the names of two German dogs of the time Sattin and Damask.[52] Names suggest the connection between expensive animals and other luxury goods, with the high status conferred by ownership of such creatures – carefully acquired or bred and prized for matching an ideal. They also indicate the enjoyment of looking and the sensation of touching soft, sometimes fluffy, fur and the affective connection related to that act.

Two portraits that have in the past been identified as Isabella d'Este with a dog are valuable for analysis of proximity, tactility, and the aesthetic qualities of Italian Renaissance lapdogs, whether we believe them to be images of Isabella or not. The first is Lorenzo Costa's *Portrait of a Lady with a Lapdog* (*c*.1500–5) (Figure 2.3). This portrait shows the sitter's fingers in the long, silky fur of a Melitaean-type dog, what we would now class as bichon. Soft, like the toy spaniel, these dogs have white fur, longer if of the Maltese type

Figure 2.3 Lorenzo Costa, *Portrait of a Lady with a Lapdog* (*c*.1500–5), Royal Collection Trust / © Her Majesty Queen Elizabeth II 2017.

(as here), shorter if of the Bolognese (as seen below). This image has been associated with Isabella d'Este, possibly with her famous Aura. The sitter is probably too young to be the thirty-year-old Isabella and may be her daughter Eleonora, and it is too early to show Aura. In any case, Aura may have been this kind of Maltese, and tributes to her tell how she was caressed in this way.[53] We can certainly imagine this dog turning round to indulge in Aura's habit of nibbling on fingers, with contact that may meld Konecki's categories of the playful touch with the affectionate. The enjoyment to be had in such games is described in a 1586 text by Annibal Guasco in which he gives advice to his daughter Lavinia on how to conduct herself at court. The author advises Lavinia on the limits of jesting:

> let your quips be like those playful nips that your little dog is wont to give you with [her] teeth, which, although applied to your flesh, nevertheless never do you any real hurt but, rather, afford you pleasure through feeling yourself harmlessly gripped by the teeth of that dear little creature.[54]

The difference between endearing nips such as these (or Aura's to Isabella) and bites that actually hurt lies in interspecies affection and communication. The nips mean something on both sides of the species boundary and must be judged as playful and not painful. Mutual physical interactions such as these, and the stroking and other games that may precede and follow them, are predicated on proximity and are at the heart of the human–companion animal relationship.

Class comes into play with these behavioural and aesthetic considerations, with ideas that softness, delicateness, and grace accompanied by good manners were the preserve of the nobility. Future research might usefully investigate the tension between potentially disruptive animal (or child) bodies, training notwithstanding, and the ideal orderly sensorium. How is the controlled behaviour and controlled body of the 'civilising process', what Niall Atkinson terms 'sensory discipline',[55] reconciled with animal bodies ever present in the palace?

The second image that may or may not be Isabella is the *Portrait of a Woman* now attributed to Giulio Campi (Figure 2.4). This shows a lady in Isabella's trademark *capigliara* headdress, with another dog who may enjoy chewing on

Figure 2.4 Portrait of a Woman, by Giulio Campi © Hearst Castle®/CA State Parks.

the sitter's hand. The suggestion was made in *The Burlington Magazine* in 1929 that the dog is either Aura or a successor, but with the judicious observation that 'if it is commonly difficult to estimate the age of a woman from a portrait, it is certainly far more difficult to judge that of a dog'.[56] As with the Costa portrait, we see the same attention to the proximity of dog to owner and to the act of stroking, here hair curling around a gloved finger. Interesting, too, is the juxtaposition of living fur with dead, in the presence of the marten.[57] In showing an animal being touched, the artist is challenged to paint fur that begs to be felt, and to invite the viewer to share imaginatively the sensation of luxurious touching. Such depiction in portraits indicates possession by the owner and an emotional connection between the owner and animal in the act of stroking and being stroked.

Senses, like emotions of course, have a history built upon culture, as Peter Burke has recently pointed out, to the 'period eye' we can add 'the period ear or nose',[58] or indeed hand. Although a neo-Platonic hierarchy of senses prioritised sight, touch might also be celebrated, drawing support from Aristotle, as for instance in the work of Mario Equicola, Isabella's secretary.[59] In addition to appearance and characteristics related to touch and tactility, those indicative of proximity in sound and smell are also important in Renaissance companion animal aesthetics and sensory appeal. There is evidence to suggest the perfuming of dogs at court,[60] and Isabella's dogs famously proclaimed her presence with their yaps.[61] Isabella's apartments, and in particular her showpiece *grotta* and *studiolo* collection spaces, can be valuably considered as *sensoria* in which Isabella was a connoisseur of finely tuned senses,[62] and in which animals played significant roles through, for instance, the barking, or soft and perfumed bodies, or shimmering coats aforementioned, or the sound and beauty of parrots in finely wrought imported cages.[63]

Amid these Renaissance worldviews and strange, culturally defined human sensory landscapes, we should not forget animal sensescapes, yet more mysterious, including olfactory universes. We might ask how the sensory environment of the court impacted upon its animal inhabitants and how the sensory experiences and engagement of these animals shaped both their worlds and human worlds. It may be valuable here to reflect on what can be revealed at the point at which attention to the history of the senses, the so-called 'sensuous turn',[64] collides with the related 'emotional turn' and with the 'animal turn'. This calls for focus not only on men and women but also a more-than-human perspective. For this, the scholar is required to attempt to imagine and understand animal senses and emotions and the interior lives of other species, while historicising sensation and emotion and the effects of these on physical and psychological wellbeing of both sides of the human–animal relation. Emotion and senses have biological, neurological bases alongside a history. The emotions and senses of animals whose species are bred (or captured) according to and as products for humans must also be influenced by a combination of nature and human culture. In 'Milking Other Men's Beasts', Erica Fudge asks whether the perspective offered by Temple Grandin can illuminate the world of the early modern cow, notwithstanding changes in cow bodies in the intervening centuries.[65] Following this lead, these issues can be unpicked further for both the human and animal perspectives by more research and interdisciplinary conversation.

Another area for questioning lies in the implications for health and well-being of proximity, affection, and sensory relationships with companion animals in the past. Does historical evidence of people living side by side with animals, face to face, stroking or touching them, show attachment or pleasure in physical contact or something else like wanting to warm their bodies or be rid of fleas, or all of these? Might it show an intuitive understanding of increased wellbeing through affiliative bonding with animals?

On this, as other matters, it is fruitful to be open to dialogue between historical animal studies and medical humanities as well as the social sciences, life sciences, and human and animal medicine. Modern studies suggest that proximity to companion animals is therapeutic to humans, with an impressive array of psychological and physiological effects: increasing self-esteem, alleviating stress, releasing dopamine and endorphins, improving immune system function and pain management, and lowering blood pressure, for instance.[66] The realms of modern science and social science may help us gain insight into mechanisms of companionate bonding and psychological benefits of pet keeping that can illuminate past human-companion animal relationships, even if not conceived of in the same way in Renaissance Italy. Recent studies seek to show, for example, that the canine brain and human brain share similarities that help us understand each other and that dogs read human faces;[67] and research identifies the importance of companionship, affinity, care-giving and mutual affection.[68] Dogs of the bichon group, particularly the Bolognese, are often trained today as working therapy dogs to assist human practitioners. The dogs' temperament and breed history make them excellent at soothing humans and helping their psychological state, and this may be reflective of the breed's beginnings in elite domestic environments. Recognising ourselves as products of our own cultural concerns, it may nevertheless refresh our view of past relationships to bring recent studies to bear on historical sources that suggest that rulers found companion animals important in alleviating a sense of isolation and in providing a trusted friend in a Machiavellian world, or on images such as the Passerotti portrait (Figure 2.5) with a dog reading the face of his human.

There is also ample evidence in modern research of the contribution of sensory factors in delight in the human-animal bond – seeing and interacting with animals who are often still bred or selected for luxurious coats and neotenised features. Recent scientific study finds not only the presence of companion animals to be beneficial to human wellbeing but particularly the act of touching and stroking them.[69] It should be noted that few of these studies consider the benefits for animals of being touched by humans. One study that did concluded that when adults petted a dog, there was an increase in oxytocin, beta-endorphin, prolactin, beta-phenylethylamine, and dopamine levels in both human and canine subjects, therefore suggesting

Figure 2.5 Passerotti, *Portrait of a Man with a Dog* (1585–7), Rome, Pinacoteca Capitolina Museums 2.

beneficial effects for both parties.[70] More studies would be welcome on the effects of human–animal interactions that look committedly at both sides of the equation. Furthermore, this research reflects a phenomenon pointed to by Karen Raber in *Animal Bodies, Renaissance Culture* in relation to elephant cognition and culture: 'evidence, particularly statistical evidence wedded to advances in neurological science, is king'.[71] We want science to confirm what many people, in cultures past and present, already know: in this case, that being in the company of a companion animal who is relaxed and content can lead to human relaxation and contentment, and these things may well be good for health. All in all, to the historian, oxytocin, like the biology of sensory perception or the neuroimaging of brain regions connected to affective functions, may be of interest but is ultimately of lesser concern. If a human or animal body responds here and now to positive interspecies interaction in a certain way under experimental conditions, it does not perhaps matter if the same hormone was released, in the hypothalamus of either woman or dog,

500 years ago in Italy. In any case, how could we know? The acculturated body, senses, and emotions must be read of their own time and place,[72] but if the findings of modern science, medicine, or social science can help our historical questions speak to the world around us – across the academy and beyond; shape some questions usefully asked of the past; and ultimately aid in the development of a more informed view of human–animal relationships then and now, then such engagement must be embraced.

Bereavement

A factor that studies have identified as damaging to human wellbeing is that of pet bereavement (more research on the effects of owner bereavement on the animal would be of value).[73] While ever mindful of cultural difference across space and time, the historical record shows us the depth of feeling for lost companion animals in the past. On this theme, the story of the *animalino* quickly takes a sad turn. After little over two weeks in Mantua, the *animalino* sickened and, despite medical attention, died.[74]

It would have been fascinating, had the *animalino* lived, to watch Isabella's relationship with this animal become established, calling him by a name, for instance. His is an interesting case. On one hand, the *animalino* was a commodity; he had been traded, gifted, and transported, and he was valued for certain qualities that were the same as other luxury objects: rarity, delicacy, aesthetic charms of beauty and touch. Beyond this, he was a living creature, pampered and babied. Even though he had been in Isabella's possession only a short time, her emotional attachment to the cat appears to have been strong. For Isabella the shopper and the pet-keeper, his loss was a double blow. The preciousness of the *animalino* lay in part in his tininess, his softness, his adorability; in part in his unique status as an exotic animal; and in his donor. It is telling perhaps that Isabella's thoughts after communicating the death of the *animalino* turned immediately to concerns that Federico stay in good health. Isabella wrote to her son:

> The beautiful little cat that you sent, becoming so seriously ill that no good remedy could help him, finally passed away. This must have been caused by the change of air and his delicate nature, as he did not lack a single thing nor good care both on his journey and here. You make sure to keep yourself healthy and *bene vale*.[75]

The Gonzaga often went to great lengths to save favourite companion animals, enlisting the skills of their most respected court physicians.[76] Isabella stressed that the *animalino* was cared for scrupulously, but nevertheless something went wrong. Had he been delivered with, or by, a more skilled handler, the availability of knowledge about the animal may have proved vital.[77] What intervention did the *animalino* receive and did his exoticism and unknowability prove his downfall? If the *animalino* was a manul, for instance, keeping Pallas' cats healthy in captivity is not easy, as their immune systems are exposed to viruses not encountered in their high-altitude natural habitat.[78] We might concur with Isabella in blaming, then, 'the change of air and his delicate nature'. If the *animalino* was sickening even on his arrival in Mantua, which may be a reasonable assumption with his death so soon after, then his tractability may have been in part due to weakness and ill health. Ultimately evaluating the wellbeing of the *animalino* during his time in Mantua remains difficult.

For the human side of this relationship we see the negative effects of pet bereavement. I have not yet uncovered a record for a funeral or tomb for the still-unnamed *animalino* but I hold out hope of finding one. Two years earlier, in 1510, Isabella had organised a funeral – attended by many animals – for a favourite cat, Martino.[79] In 1511, Isabella was mourning another favoured animal after the accidental death of Aura. As Isabella was readying her entourage to depart after a visit, Aura was racing around fighting with Isabella's other dog Mamia over the affections of a third dog when she lost her footing, fell over 10 metres from a small terrace, and died.[80] Isabella grieved conspicuously for Aura, crying at dinner and sighing. This is striking, as Isabella was keen to maintain her composure with significant human losses in her life. She allowed herself to put aside the equanimity to which she aspired to mourn 'the most beautiful and pleasing little dog there ever was'.[81]

Famously, Isabella raised a tomb to Aura, and the dog's death provoked a competitive flurry of literary tributes in Latin and the vernacular by poets across Italy.[82] We see a pattern in the Gonzaga response to the death of favoured animals, with funeral rituals, elegies, and epitaphs.[83] These follow classical precedents and are fashionable humanist exercises, but I do not agree with scholars who have written off a genuine emotional component to these or recognition of this by the humanists involved.[84] Scholarship has tended to consider these tributes from the literary perspective and in line with Isabella's scrupulous self-fashioning as a cultural patron.[85] Beyond this, Isabella's grief is also apparent. Such devastation at the loss of a companion animal was not rare in Renaissance Mantua. Lodovico Gonzaga had tombs erected to Bellina and Rubino. When Rubino suffered from tumours in old age, he was offered the finest cooked food and raw meat and was attended by a physician, Lodovico ordering that no effort be spared to save him. When all failed, Rubino was buried with a tomb and fine epitaph in a spot that Lodovico could see from the *Camera degli Sposi* in the palace where he and Rubino had enjoyed spending time together.[86]

Isabella's son Federico Gonzaga in his adulthood also memorialised his favourite dogs with tombs, such as the example in the *giardino segreto* of the Palazzo Te (Figure 2.6). This stucco tomb is now attributed to Giovan Battista Mantovano (*c*.1531) but has previously been put forward as the monument designed by Giulio Romano for a dog who, like Lodovico Gonzaga's Bellina, had died giving birth. This dog may have been the Viola whose epitaph was recorded, along with many other Mantuan examples, by Ulisse Aldrovandi in his *De quadrupedibus digitatis viviparis* (1645).[87]

Figure 2.6 Giovan Battista Mantovano, monument to a *cagnolina* of Federico Gonzaga, stucco, Appartamento del Giardino Segreto, Palazzo Te, Mantua
(*c.*1531).

We similarly do not know the name of the dog in Titian's portrait of Federico of 1529 (Figure 2.7). Again, this may be Viola – possibly portrayed post-
humously – or one of her offspring or a successor. It has been suggested that
the dog in this portrait plays a purely or predominantly symbolic role, representing Federico's commitment to conjugal fidelity as a prospective husband
or the verisimilitude of Federico's portrait.[88] However, along with potential
symbolism, there is more going on in this painting. This is a portrait of the
stroking of an actual dog. As well as control, this image gives an example of
Konecki's communicative touching of a pet, 'characterised by positive affection', par excellence. Federico's hand sinks into the dog's fur, and she reaches
out to return his touch. The artist makes evident her softness, her tactile appeal, and the strength of the bond between the man and the dog.[89]

Figure 2.7 Federigo [Federico] Gonzaga, Duke of Mantua from 1530, *c.*1525–30 (oil on panel), Titian (Tiziano Vecellio) (*c.*1488–1576) / Prado, Madrid, Spain / Bridgeman Images.

What is Federico doing portraying himself with this small dog? His, and other male rulers', ownership of and affection for these dogs is not unusual, as indicated with the example of Federico's great-grandfather Lodovico Gonzaga above. There was a backlash against extreme fondness of ladies or men for little dogs.[90] For instance, Henri III of France was criticised for his excessive love of small dogs and giving them an expensive retinue.[91] But Federico, like Lodovico before him, seems to have attracted no censure for this. His relationship with this companion dog is played out here with elegance. It shows an easy courtly authority and perhaps a marriageability, through not just the symbol but also the affection. Here is a real relationship with a real dog.

Conclusion

To take a step forward from an animal depicted in an image, glimpsed in documents, or commemorated in stone, it is important for research to incorporate the materiality of live animal bodies and to consider the effect of movement and senses. A real dog, Titian's is a portrait of a dog who is cared for, interacting in meaningful ways with people in the past, and a dog who cares in return for her owner.

In the haptic relationship of owner and animal, stroking is an art, an act of interspecies communication through touch (accompanied by sight, smell and perhaps by sound and taste). A stroke down the back, tickle on the tummy, scratch behind the ear, pat on the head, or a reassuring (perhaps controlling) hand to say 'I am here'. These are in part instinctive as ways for a human to communicate across species, but knowing how to touch each animal as an individual, safely and in ways to which the animal responds well, takes time to learn as well as trust on both sides. On the animal side of the equation, the same may be true of a cat's rubbing against a leg or – of the dogs met here – of playful nips, a paw outstretched, or a darting lick (for which Passerotti's dog is surely poised).

This chapter has argued that the relationships between Mantuan rulers and their companion animals were based on markers of power, status, and ownership, but also on a close proximity, aesthetics, tactility, and sensory delight, as well as on an affection reflected in attachment in life, anxiety in illness, and grief and memorialisation in death. In seeking to understand the human emotions and implications for human wellbeing in these relationships, we must continue to ask: what about the animal? To assess the effects of their experiences on the wellbeing of animals in circumstances often beyond their choosing, how might we know whether Rubino, Bellina, and Aura were contented dogs? The sources seem to show they were attached to their owners, lived close to them, and were assiduously cared for, and in the case of Rubino came home by choice when lost, but beyond that? How can we judge animal affection? And what of the *animalino*? The question is even thornier for an animal whose species is not certain, whose health deteriorated despite what was deemed the best available care. How did he feel to be in a sleeve? It is clearly important for us when we study the human and animal past to take these questions into account, though it is not an easy endeavour. To try to understand the *Umwelt* of another species – or indeed the lives of others more generally[92] – we must engage in the imaginative. For this task in particular, by necessity, we must take part also to some extent in the anthropomorphic (because we have no other way) – with the hope that this be less by way of patronising and more in the spirit of empathy and a respectful desire to make some sense of another's world. We are required to attempt a conceptual leap across the species boundary.

In our efforts to understand we may decide, then, to leave our comfort zones and collaborate across other boundaries to work with puzzling species such as sociologists, ethologists, vets, zooarchaeologists – maintaining commitment to the rigours of our own disciplines, alongside an openness to the opportunities offered by others. I argue, for instance, that the paradigms of research into the physiological and psychological effects of the human-companion animal relationship may move historians to look with new eyes at the ways these effects were conceptualised and lived in the past.

We may collaborate too, of course, with the animals that we know best, not only disciplinary colleagues of our own species but also those creatures of other species with whom we may share proximity, sense space, and companionship. To end as we began, by borrowing a phrase from Donna Haraway, following Claude Lévi-Strauss, '[dogs] are not here just to think with. They are here to live with.'[93] What did, and does, it mean to share our homes with animal bodies? Animals that smell, bark, yip, run, growl, nip, curl up, eat, play, lick, touch, and are touched? What does it mean to feel an emotional connection with an animal in the past, or now?

Notes

1 I would like to thank Andrew Wells and Erica Fudge for their valuable comments on this chapter as well as the Leverhulme Trust, Wellcome Trust, and University of Glasgow for kindly funding research, meetings, and opportunities for interdisciplinary conversations on this topic. I am, as always, grateful to the staff of the Archivio di Stato in Mantua for their kindness and assistance.

2 The term *animalino* denotes the animal's diminutive size and is found often to refer to companion animals in Renaissance Italy, not always cats or even dogs. See note 54 below. The term may also, in the case of Isabella's *animalino,* suggest the unknown or unusual species of the tiny animal. The related term *animaletto* is used by the Medici, for instance, to refer generically to little animals as well as to small exotic imports. See, for example, Francesco I di Cosimo de' Medici to Ferdinand von Habsburg, Florence, 9 September 1581, sending 'un animaletto venutomi dell'Indie che chiamano lepre di quel paese', Florence, Archivio di Stato, Mediceo del Principato, Volume 257, Folio 30.

3 A selection of this scholarship includes Kathleen Kete, *The Beast in the Boudoir: Petkeeping in Nineteenth-Century Paris* (Berkeley and Los Angeles: University of California Press, 1994); Harriet Ritvo, 'The Emergence of Modern Pet-Keeping', *Anthrozoös,* 1 (1987), 158–65; James Serpell, *In the Company of Animals* (2nd ed., Cambridge: Cambridge University Press, 1996); Ingrid H. Tague, *Animal Companions: Pets and Social Change in Eighteenth-Century Britain* (University Park: Penn State University Press, 2015) and Kathleen Walker-Meikle, *Medieval Pets* (Woodbridge: Boydell Press, 2012).

4 For 'entwinement', see Erica Fudge's foreword to this volume.

5 Erica Fudge, 'Milking Other Men's Beasts', *History and Theory,* 52/4 (2013), 13–28, p. 23.

6 Krzysztof Konecki, 'Touching and Gesture Exchange as an Element of Emotional Bond Construction. Application of Visual Sociology in the Research on Interaction between Humans and Animals', *Forum Qualitative Sozialforschung / Forum: Qualitative Social Research* 9/3 (2008) www.qualitative-research.net/index.php/fqs/article/view/1154, accessed 23 January 2014.

7 For animals' place in Renaissance sensescapes see also Karen Raber, *Animal Bodies, Renaissance Culture* (Philadelphia: University of Pennsylvania Press, 2013), 22–7.

8 Donna Haraway, *When Species Meet* (Minneapolis: University of Minnesota Press, 2008), 206. Also Yi-Fu Tuan, *Dominance and Affection: The Making of Pets* (London: Yale University Press, 2004).

9 Sarah Cockram, *Isabella d'Este and Francesco Gonzaga: Power Sharing at the Italian Renaissance Court* (Women and Gender in the Early Modern World; Farnham: Ashgate, 2013).

10 Baldesar Castiglione, *The Book of the Courtier,* trans. George Bull (London: Penguin, 1976), 310. Baldessar Castiglione, *Il Libro del Cortegiano* (Milan: BUR,

1994), Book 4, 36: 'far conviti magnifici, feste, giochi, spettacoli publici; aver gran numero di cavalli eccellenti, per utilità nella guerra, e per diletto nella pace; falconi, cani e tutte l'altre cose che s'appartengono ai piaceri de' gran signori e dei populi; come a' nostri dì avemo veduto fare il signor Francesco Gonzaga, marchese di Mantua, il quale a queste cose par più presto re d'Italia, che signor d'una città'.

11 See, for instance, Lisa Jardine and Jerry Brotton's *Global Interests: Renaissance Art between East and West* (Ithaca: Cornell University Press, 2000), 149–53, and Giancarlo Malacarne's, *Le cacce del principe: l'ars venandi nella terra dei Gonzaga* (Modena: Il Bulino, 1998), 26–7, 49, and *I signori del cielo: La falconeria a Mantova al tempo dei Gonzaga* (Mantua: Artiglio, 2003), 161–95, 258–307.

12 For Gonzaga family devices, including the *cane alano* and muzzle see David Chambers and Jane Martineau (eds), *Splendours of the Gonzaga* (London: Victoria and Albert Museum, 1982), 173–74. For Gonzaga pets see Clinio Cottafavi, 'Cani e gatti alla corte dei Gonzaga', *Il ceppo quaderno di vita fascista e di cultura* (Mantua, 1934), 8–11.

13 For example, Malacarne, *Le cacce*, 49. On surviving artefacts see, for instance, Elizabeth Wilson and Wilson Stephens, *Four Centuries of Dog Collars at Leeds Castle* (London: Leeds Castle Foundation, 1979). The material culture of companion animals can offer much on the subject.

14 Lodovico Gonzaga to his wife Barbara of Brandenburg, 30 October 1462, Archivio di Stato di Mantova (henceforth ASMn), Archivio Gonzaga (AG), busta (b.) 2097: 'Rubino è gionto questa matina [...] El ne ha facto careze assai e ne ha domandato perdonanza [for running away], e così gli habiamo perdonato'. Quoted by Rodolfo Signorini, 'A Dog Named Rubino', *Journal of the Warburg and Courtauld Institutes*, 41 (1978), 317–20, p. 317.

15 See also Ido Ben-Ami's chapter in this volume for a lion portrayed as fearful of the Ottoman sultan.

16 Pierre Belon du Mans, *L'histoire de la nature des oyseaux...* (Paris, 1555), Livre III, Chapter 2: 'comme nous tenons quelque petit chien pur compagnie, que faisons coucher sur les pieds de nostre lict pour plaisir: iceluy y avoit telles fois quelque Lion, Once, ou autre telle fiere beste, qui se faisoyent chere comme quelque animal privé es maisons des paisants'. See also Gustave Loisel, *Histoire des ménageries de l'antiquité à nos jours* (Paris: O. Diens, 1912), 264.

17 For exotic animals as expressions of princely power see also Sarah Cockram, 'Interspecies Understanding: Exotic Animals and their Handlers at the Italian Renaissance Court', *Renaissance Studies* (Special Issue: 'The Animal in Renaissance Italy', eds Stephen Bowd and Sarah Cockram), 31 (2017), 277–96. doi:10.1111/rest.12292.

18 Serpell, *In the Company of Animals,* 43.

19 To give an early modern example, the duchess of Windsor in 1737 wrote to her daughters: 'I'm very fond of my three dogs, they have all of them gratitude, wit and good sense; things very rare to be found in this country'. Serpell, *In the Company of Animals*, 53. Katharine MacDonogh explores the attachment of rulers to their pets as providers of affection and loyalty, *Reigning Cats and Dogs: A History of Pets at Court since the Renaissance* (London: St. Martin's Press, 1999), 206–7.

20 Francesco Malaguzzi Valeri, *La corte di Lodovico il Moro. La vita privata e l'arte a Milano nella seconda metà del Quattrocento* (Milan: Hoepli, 1913), i, 58.

21 See, for instance, Juliana Schiesari, *Beasts and Beauties: Animals, Gender, and Domestication in the Italian Renaissance* (Toronto: University of Toronto Press, 2010), 17–19.

22 The vast scholarship on Isabella might be approached through recent studies such as Stephen J. Campbell, *The Cabinet of Eros: Renaissance Mythological Painting and the Studiolo of Isabella d'Este* (New Haven: Yale University Press, 2006);

Cockram, *Isabella d'Este and Francesco Gonzaga*; Carolyn James, 'Machiavelli in Skirts. Isabella d'Este and Politics', in J. Broad and K. Green (eds), *Virtue, Liberty and Toleration: Political Ideas of European Women 1400–1800* (Dordrecht: Springer, 2007), 57–76; and the publications of Deanna Shemek, including her valuable edition of Isabella's letters in translation (Toronto and Tempe: Iter Academic Press and Arizona Center for Medieval and Renaissance Studies, 2017). For music see, for instance, William F. Prizer, 'Una "Virtù Molto Conveniente a Madonne": Isabella d'Este as a Musician', *The Journal of Musicology*, 17 (1999), 10–49.

23 Malacarne, *Le cacce*, Chapter 1. For Isabella's naming of two spaniel puppies to recall her famous motto *nec spe nec metu*, see Isabella to Giovanni Volta, Mantua, 14 January 1510, AG, b. 2917, Libro (L.) 210, folio (fo.) 13v: 'Volemo ch'el maschio habbi nome Metus, la femina Spes'.

24 See Mario Equicola's elegy to Aura, ASMn, Serie Autografi 10, b. 356, fo. 5v-6r.

25 Federico to Isabella, Rome, 5 June 1512, ASMn, AG, b. 2119 bis, fascicle (henceforth fasc.) XI, carta (c.) 526: 'perché io sempre penso ad far cosa che la possa piacere, ho procurato di haver il più bello gatino più delicato che sia mai venuto de Colocut portato qua da Messer Angelo del Buffalo fino de Portugallo et dice haverlo havuto da la Regina in dono, et giura che in Spagna né in Italia è sta portato altro gatino di questa sorte se non questo, di modo che è rarissimo anzi solo in Italia almeno, l'è tanto bizarro in ogni cosa che è maraviglioso, et è picolino per portare in una manica. M. Angelo è sta contento di donarmilo benché l'havesse promisso alla Duchessa di Urbino mia sorella la qual haverà patientia perché non intendo che altra che Vostra Excellentia l'habbia, e parmi che non sia venuta cosa in Roma che più debbia piacere a Lei che questo animalino delicatino et maraviglioso, vedo ben che Vostra Signoria se lo tenerà in grandissime delicie et haverà ragione perché è cosa desiderata da ciascuno ch'el vede, et il Cardinale di Ongaria ha posto in electione M. Angelo de tutti li soi cavalli che molti belli ne ha, ch'el si cappa qual vole et lo piglia per questo gatino. Io l'haveria mandato hora per Antonio cavallaro ma perch'è tanto delicatino non voria patesse sinistro nel viagio maxime non intendendo lui la natura dil animalino, et voluntieri son condesceso ad acceptar la offerta di M. Angelo che tanto amorevolmente mi ha fatto M. Angelo [sic] di mandar a posta a Mantua a portar esso gatino colui istesso che l'ha governato sempre di Spagna sin qua perché melio cognosce la natura dil animale che altri, se che tra quatro giorni credo di inviarlo et con quello di M. Angelo mandare uno cavallaro che l'uno e l'altro ni havrà cura per posserlo condur a salvamento a Vostra Excellentia (My transcription here and elsewhere expands contractions and alters punctuation when required for clarity, but retains original spelling.)

26 Isabella suffered from infestation of rodents in her rooms, and she asked the Gonzaga agent in Venice, Giorgio Brognolo, to source four Syrian cats to catch rats. The previous year rats' nests were discovered under the floorboards of Isabella's *studiolo* as tiles were being laid. Dora Thornton, *The Scholar in His Study* (London: Yale University Press, 1997), 48–49. See also Clifford M. Brown and Anna Maria Lorenzoni, *Isabella d'Este and Lorenzo da Pavia: Documents for the History of Art and Culture in Renaissance Mantua* (Geneva: Droz, 1982), 245–46. For *gatti de spagna* for fur, see Isabella to Gian Battista Vicemala, Mantua, 24 January 1506, AG, b. 2994, L. 18, c. 64r.

27 See other examples in Naomi Sykes, *Beastly Questions. Animal Answers to Archaeological Issues* (London: Bloomsbury, 2014), Chapter 7, and Richard Thomas, 'Perceptions versus Reality: Changing Attitudes towards Pets in Medieval and Post-medieval England', in Aleksander Pluskowski (ed.), *Just Skin and Bones? New Perspectives on Human-Animal Relations in the Historic Past* (Oxford: Archaeopress, 2005), 95–105.

28   A long-list of potential species may also include the Leopard Cat (*Prionailurus bengalensis*), Asiatic wildcat (*Felis silvestris ornata*), the sand cat (*Felis margarita*) or the flat-headed cat (*Prionailurus planiceps*) My thanks to Andrew Kitchener, Principal Curator of Vertebrates at the National Museums Scotland, for advice on identification of the *animalino*'s species.

29   Isabella to Federico, Mantua, 16 June 1512, AG, b. 2119 bis, fasc. X.1, c. 436: 'Aspettamo il gattino [...] in desiderio incredibile per essere di la novità, belleza et delicatura che ni scrivi'.

30   Isabella to Federico, Mantua, 29 July 1512, AG, b. 2119 bis, fasc. X.2, c. 456: 'Li gatini che ne dici de mandarne per Federico Catanio ne serano molto grati', and Mantua, 20 August 1512, c. 462: 'De li tri gatini che per esso Federico [Cattaneo] ne scrivi mandarne uno havemo inteso per la via esser manchato, dilche havemo sentito pocho dispiacere poi che quel bello et singulare animalino [è] condutto salvo insieme con l'altro gatino. Teneremo tra le altre nostre cose care lo Animalino Carissimo sì perché l'è de summa beleza e de nova forma presso ogniuno de qua come maxime per havernelo tu mandato. Siché lo conservaremo et attenderemoli de nostra mane et per tua memoria et per nostro non mediocre spasso'.

31   Giovanni Giacomo Calandra to Federico Gonzaga, Mantua, 14 August 1512, AG, b. 2485, cc. 148r–149r: 'Non potrei scrivere con quanta allegreza Sua Excellentia [Isabella] accettò el nobilissimo presente di Vostra Signoria la quale di essere certa che la non poteva mandare cosa che fosse più grata per la belleza né più mirabile per la sua rarità di quel bello animalino. Fu forza che esso Messer Federico [Cattaneo] differisse molto in longo a fare la sua ambassata fin che Madamma se sacciasse alquanto in mirare et laudare quel animalino delicato, el quale fin qui non ha altro allogiamento che la manica et seno di Sua Excellentia [...] Uno di gattini, cioè il maschio non ha potuto giungere vivo a Mantua, che è ben rincresciuto forte a Madamma, ma la gentileza del piccolino recompensa el danno di quello. Madonna gattina sta in la camera di Madamma in delicie et feste di ogniuno per essere così piacevole ma beato chi può toccare el bellino. [...] La Signoria Vostra non potria credere la cura che ha Madamma continuamente che quel animalino repossi mollemente et mangi cose che gli piacia, ma non se sacia di basarlo et festeggiarlo con le più tenere paroletti del mondo. Lo amore de la sua Aura era niente appresso questo. La gli fa fare una lettica bellissima da alloggiarlo ma vi starà poco, perché [...] lo alloggiamento suo ha ad esse ella istessa. La morte del gattino è stata perché Santo, famiglio di M. Mario, amalò a Siena et ivi restò, onde el gattino bisognò cavalcare un cavallo da sua posta menato a mano el quale stazandose ad una colonna in una hostaria lo amazò.' Federico Cattaneo to Federico Gonzaga, Mantua, 15 August 1512, b. 2485, c. 279r-v. This letter opens with explanations and excuses for the accidental death of the black male cat on the journey, before continuing: 'lo gatino rosso et la gatina negra insieme como li altri cosi li presentai ala Illustrissima Madonna vostra matre per parte di Vostra Signoria, dove Sua Signoria ne fece tanta festa del gatino rosso che più non poteria scrivere a Vostra Signoria de sorte pocho Sua Signoria curò quel che era morto tanto alegreza fece [...] di continuo Sua Signoria lo porta [the animalino] in la manega sino a Vespero [...] lo Illustrissimo Signore vostro patre essendo a [...] Santo Petro lo fece portare alui et como lui lo ebbe in brazo [...] tuta quela giente steteno admirativi a vedere simele animalle.'

32   Serpell, *In the Company of Animals*, 50, 53.

33   See, for instance, Evelyn Welch, 'New, Old and Second-hand Culture. The Case of the Renaissance Sleeve', in Gabriele Neher and Rupert Shepherd (eds), *Revaluing Renaissance Art* (Aldershot: Ashgate, 2000), 101–119.

34 For fur-lined sleeves see Patricia Lurati, '"To dust the pelisse": The Erotic Side of Fur in Italian Renaissance Art', *Renaissance Studies*, 31 (2017), 240–60. doi:10.1111/rest.12293. It is unlikely in this case, however, that the *animalino* rested in fur-lined sleeves in the heat of a Mantuan August.

35 Castiglione, *Il Libro del Cortegiano*, I, 26.

36 Evidence from the case of the *animalino* is able to begin to fill a gap identified by Karen Raber: 'petting cats simply does not figure in early modern texts or cultural artifacts, although we must assume it happened'. Bacchiacca's ladies with cats might also dispute Raber's assertion that 'not until the early eighteenth century were cats in paintings depicted as lap cats, being held – and presumably petted – by their owners'. Raber, *Animal Bodies, Renaissance Culture*, 25.

37 Thomas Tuohy, *Herculean Ferrara* (Cambridge: Cambridge University Press, 1996), 84.

38 Francesco Trevisano to Isabella d'Este, Venice, June 1501, AG, b. 1439: having 'una gata suriana per dover mandar a donar a Vostra Illustrissima Signoria, è intervenuto che, per malla fortuna essendo in amore […] saltò zoso di balchoni, quamvix altissimi, de una de le camare de la mia caxa dove l'era serrata'. Quoted by Brown and Lorenzoni, *Isabella d'Este and Lorenzo da Pavia*, 245.

39 Juliana Barnes, *Boke of Saint Albans* (1486), quoted by Schiesari, *Beasts and Beauties*, 19. On fleas see Raber, *Animal Bodies, Renaissance Culture*, 25–27.

40 Abraham Fleming's rendering of Caius's *De Canibus Britannicis* states that: 'litle dogs are good to asswage the sicknesse of the stomacke being oftentimes thervnto applyed as a plaster preseruatiue, or borne in the bosom of the diseased and weake person, which effect is performed by theyr moderate heate. Moreouer the disease and sicknesse chaungeth his place and entreth […] into the dogge […] these kinde of dogges sometimes fall sicke, and sometime die'. Despite this service, Caius is also scathing about excessive affection and proximity to lapdogs, see note 90 below. Johannes Caius, *Of Englishe Dogges, the Diversities, the Names, the Natures and the Properties*, trans. Abraham Fleming (London, 1576), 20–22.

41 See Cockram, *Isabella d'Este and Francesco Gonzaga*, Chapter 6.

42 The Gonzaga distributed the offspring of their companion animals as markers of favour to courtiers, following the pattern in all high-status commodities (e.g. rooms, clothing, food), that the finest be for the ruling family, then trickle down the court hierarchy. For instance, the courtier Alfonso Facino, 'trinzante' refers to 'mio cagnolo mi ha donato madama', to Federico Gonzaga, Mantua, 21 November 1510, AG, b. 2479, fasc. X, c. 261. The Gonzaga also sought out family relations of their best dogs, and in this manner favour might be won by a donor, for instance Lodovico was pleased to receive a descendent of Rubino as a gift after the dog's death. Signorini, 'A Dog Named Rubino', 320.

43 Giovanni Giacomo Calandra to Federico Gonzaga, Mantua, 21 April 1511, AG, b. 2482, c. 111r-v: 'La Fanina gli dì passati partorì tre figlioli: una cagnolina viva, dui maschi morti. La cagnolina hebbe la Signora donna Hippolyta, et gli pose nome Fratilla. Poi la damma n'ha fatti quattro. La illustrissima Signora vostra madre racordandose de Vostra Signoria me ne ha dato uno per Lei, il più bello et piacevolino dil mondo, rossetto sfazato de peza biancha in mezzo la fronte. In mezzo el collo un'altra ha, il collo quasi tutto intorniato de circulo bianco, gli piedi tutti balzani. La punta de la coda bianca tutto allegro con bel musino. Io con consentimento della predetta Madama [Isabella] gli ho posto nome Zephyro alludendo ad Aura de Madama. Credo che la Signoria Vostra haverà un bel cagnolo. Io non gli manco de diligentia per allevarlo ben accostumato et piacevole. Zorzino ha havuta una sorella, M. Benedetto Lacioso l'altra, M. Francesco Cantelmo el terzo, ma quel de Vostra Excellentia è il più bello de tutti, com'era debito.'

44 Mamia and Zephyro's offspring include the conjoined puppies kept by Isabella after their stillbirth as an item of scientific interest and display. Dario Franchini, *La scienza a corte. Collezionismo eclettico natura e immagine a Mantova fra Rinascimento e Manierismo* (Rome: Bulzoni, 1979), 108.

45 Isabella to Teodora Angelini in Ferrara, Porto, 29 June 1493, AG, b. 2991, L. 3, fo. 69v: 'la sua cagnola negra et quello bello gatto che la teneva in delicie'.

46 I give an example elsewhere of delight derived from the unexpected behaviour of a Ferrarese cheetah on a hunt with the king of France. Cockram, 'Interspecies Understanding'.

47 Signorini, 'A Dog Named Rubino', 317.

48 On this subject see, for instance, Vicki Hearne, *Adam's Task: Calling Animals by Name* (New York: Knopf, 1982) and Donna Haraway, *The Companion Species Manifesto: Dogs, People, and Significant Otherness* (Chicago: Prickly Paradigm Press, 2003).

49 For instance, Isabella sought from Lorenzo da Pavia a cat that was 'well marked and beautiful'. Mantua, 31 November 1498, AG, b. 2992, L. 9, fo. 90v: 'ben machiato et bello'. See Brown and Lorenzoni, *Isabella d'Este and Lorenzo da Pavia*, 47–48.

50 Schiesari, *Beasts and Beauties*, 30.

51 For the courtly aesthetics of lustre in relation to 'the materiality of signorial bodies', see Timothy McCall, 'Brilliant Bodies: Material Culture and the Adornment of Men in North Italy's Quattrocento Courts', *I Tatti Studies in the Italian Renaissance*, 16 (2013), 445–90.

52 Walker-Meikle, *Medieval Pets*, 18.

53 Walker-Meikle, *Medieval Pets*, 106–7.

54 Annibal Guasco, *Discourse to Lady Lavinia, His Daughter: Concerning the Manner in Which She Should Conduct Herself When Going to Court as Lady-in-waiting to the Most Serene Infanta Lady Caterina, Duchess of Savoy*, trans. Peggy Osborn (London: University of Chicago Press, 2003), 88. Guasco, *Ragionamento a D. Lavinia sua figliuola, della maniera del governarsi ella in corte, andando per Dama alla Serenissima Infante D. Caterina, Duchessa di Savoia* (Turin, 1586), 27v: 'mi soccorre qui un bello esempio molto a proposito ch'io vorrei, che i tuoi motti fossero simili a quegli scherzi che suole con esso teco la tua cagnuolina co'i denti fare, i quali tutto che ti applichi alle carni non pertanto non ti punge mai al vivo: anzi godi tu di sentirti da quel dolce animalino senza offesa coi denti strignere'. The equation of quips to bites is made also by Giovanni Della Casa in Chapter 20 of his *Galateo* (Venice, 1558), drawing on Boccaccio, although here dog bites hurt.

55 Norbert Elias, *The Civilising Process*, trans. E. Jephcott (Oxford: Oxford University Press, 1994); Niall Atkinson, 'the Social Life of the Senses: Architecture, Food, and Manners' in Herman Roodenburg (ed.), *A Cultural History of the Senses in the Renaissance* (London: Bloomsbury, 2014), 19–42, pp. 28–29.

56 Georg Martin Richter, 'The Portrait of Isabella d'Este by Cavazzola', *The Burlington Magazine for Connoisseurs*, 54 (1929), 85–92, p. 92.

57 For the *zibellino* see Jacqueline Musacchio, 'Weasels and Pregnancy in Renaissance Italy', *Renaissance Studies*, 15 (2001), 172–87.

58 Peter Burke, 'Urban Sensations: Attractive and Repulsive', in Herman Roodenburg (ed.), *A Cultural History of the Senses in the Renaissance* (London: Bloomsbury, 2014), 43–60, p. 58. For the 'period eye', see esp. Michael Blaxandall, *Painting and Experience in Fifteenth-Century Italy* (Oxford University Press, 1972).

59 Mario Equicola, *De natura d'amore* (Venice, 1525). See also Atkinson, 'the Social Life of the Senses', 40–41.

60 For the perfuming of lapdogs, see Constance Classen, David Howes, and Anthony Synnott, *Aroma: the Cultural History of Smell* (London: Routledge, 1994), 71. For

Isabella's perfumes, Alessandro Luzio and Rodolfo Renier, 'Il lusso di Isabella d'Este', *Nuova Antologia*, 147 (1896), 441–69; 148 (1896), 294–324; 149 (1896), 261–86, 666–88, pp. 676–80.

61 Matteo Bandello, *Novelle* (Lucca, 1554), Part 1, Novella 30: 'si sentirono i cagnoletti abbaiare; segno che madama era venuta fuori'.

62 I thank Deanna Shemek for encouraging me to think of Isabella's rooms as *sensoria*. Also Geraldine A. Johnson, 'In the Hand of the Beholder: Isabella d'Este and the Sensory Allure of Sculpture', in Alice Sanger and Siv Tove Kulbrandstad Walker (eds), *Sense and the Senses in Early Modern Art and Culture Practice* (Farnham: Ashgate, 2012), 183–97.

63 In relation to parrots, in 1505 Isabella sent to Venice, or if necessary Cairo, for large white parrots to present to the queen of France in response to a gift of dogs. Isabella seems to have felt that these parrots would have been kept in a study, thus making the presence of such birds in the paintings in her own famous *studiolo* particularly apt, and likely portraying some living creatures nearby. In 1492, Isabella wrote to Giorgio Brognolo to remind him to get her a parrot cage, and the inventory of her *grotta* includes two cages: one in Turkish style, the other Moorish. Isabella to Giorgio Brognolo, Mantua, 25 April 1492, AG, b. 2991, L. 2, fo. 12r. Daniela Ferrari, 'L'inventario delle gioie' in Daniele Bini (ed.), *Isabella d'Este: La primadonna del Rinascimento* (Modena: Il Bulino, 2001), 21–44, p. 31, p. 33. Isabella's father also kept parrots. Tuohy, *Herculean Ferrara*, 60, 87. See also Heather Dalton, 'A Sulphur-Crested Cockatoo in Fifteenth-Century Mantua: Rethinking Symbols of Sanctity and Patterns of Trade', *Renaissance Studies*, 28 (2014), 676–94.

64 Holly Dugan, 'The Senses in Literature: Renaissance Poetry and the Paradox of Perception' in Herman Roodenburg (ed.), *A Cultural History of the Senses in the Renaissance* (London: Bloomsbury, 2014), 149–68, p. 151.

65 Fudge, 'Milking Other Men's Beasts', 25.

66 Sandra Barker and Aaron R. Wolen, 'The Benefits of Human–Companion Animal Interaction: A Review', *Journal of Veterinary Medical Education*, 35 (2008), 487–95; June McNicholas et al., 'Pet Ownership and Human Health: A Brief Review of Evidence and Issues', *British Medical Journal*, 331 (2005), 1252–54; Daniel Mills and Sophie Hall, 'Animal-Assisted Interventions: Making Better Use of the Human-Animal Bond', *Veterinary Record*, 174 (2014), 269–73. A recent article by an international team suggests that activation of the oxytocinergic system may underlie the beneficial effects of the relationship between humans and companion animals as found across studies. Andrea Beetz et al., 'Psychosocial and Psychophysiological Effects of Human-Animal Interactions: The Possible Role of Oxytocin', *Frontiers in Psychology*, 3 (2012) doi:10.3389/fpsyg.2012.00234.

67 Corsin A. Müller et al., 'Dogs Can Discriminate Emotional Expressions of Human Faces', *Current Biology*, 25 (2015), 601–605.

68 Mills and Hall, 'Animal-Assisted Interventions', 270. See also Henri Julius et al., *Attachment to Pets: An Integrative View of Human-Animal Relationships with Implications for Therapeutic Practice* (Cambridge: Hogrefe, 2012).

69 For an overview, see Beetz et al., 'Psychosocial and Psychophysiological Effects', 3–6.

70 J.S.J. Odendaal and R.A. Meintjes, 'Neurophysiological Correlates of Affiliative Behaviour between Humans and Dogs', *The Veterinary Journal*, 165 (2003), 296–301.

71 Raber, *Animal Bodies, Renaissance Culture*, 186.

72 For more on these ideas, in relation to the contentious application of mirror neurons to Renaissance viewing of art and the dangers of universalising, see Herman

Roodenburg, 'Introduction: Entering the Sensory Worlds of the Renaissance', in id. (ed.), *A Cultural History of the Senses* (London: Bloomsbury, 2014), 1–17, pp. 16–17.

73 See, for instance, Bruce S. Sharkin and Donna Knox, 'Pet Loss: Issues and Implications for the Psychologist', *Professional Psychology: Research and Practice*, 34 (2003), 414–21.

74 Federico procured the *animalino* in June; he arrived in Mantua on 13 August and died around 1 September. Amico Maria della Torre to Federico Gonzaga, Mantua, 15 September 1512, ASMn, AG, b. 2485, c. 73r: 'Lo animalino che mandò V. S. insieme cum li gattini a la Ill.ma M.a vostra matre ne li giorni passati è morto già forsi quindeci giorni passati'.

75 Isabella to Federico, Mantua, 3 September 1512, AG, b. 2119 bis, fasc. X.2, c. 465: 'non senza nostro et tuo dispiacere intenderai el bello gatino el quale ne mandasti essersi infirmato tanto gravemente che non giovandoli alcuno studioso remedio fattoli, finalmente essere manchato. Né si pò credere che altra causa li sia se non el mutare de l'aere et la delicata natura sua, per non esserli manchato cosa alcuna ni bon governo sì per la via come qua. Tu attenderai ad mantenerte sano et bene vale.'

76 Signorini, 'A Dog Named Rubino', 317. Physicians might also find themselves carrying out post mortem dissections to determine cause of death, as in the case of the suspected poisoning of one of Francesco Gonzaga's dogs in 1513. Alessandro Luzio and Rodolfo Renier, *La coltura e le relazioni letterarie di Isabella d'Este Gonzaga*, ed. Simone Albonico (Milan: Sylvestre Bonnard, 2005), 32.

77 See Cockram, 'Interspecies Understanding' for an example of another small cat delivered without the requisite information (in this case about his diet) and for evidence of the value of the expert handler at court. For the planned and actual delivery of the *animalino*, see notes 25 and 31 above.

78 See Irina Alekseicheva (Moscow Zoo), 'Some Statistics on the Breeding of Pallas' cats in the European Population', Pallas Cat Study and Conservation Program www.savemanul.org/eng/articles/, accessed 26 June 2016. A keeper at an animal park in England recently suggested to the BBC, for instance, that with the immune systems of Pallas' cats they would expect only one or two kittens in a litter of seven to survive. http://news.bbc.co.uk/1/hi/england/7609351.stm, accessed 26 June 2016.

79 Battista Scalona to Federico Gonzaga, Mantua, 28 November 1510, ASMn, AG, b. 2479, c. 270r and Alfonso Facino to Federico Gonzaga, Mantua, 21 November 1510, AG, b. 2479, fasc. X, c. 261r-v. See also Nicoletta Ilaria Barbieri, 'Cultura letteraria intorno a Federico Gonzaga, primo Duca di Mantova' (tesi di dottorato, Università Cattolica del Sacro Cuore, Milan, 2012), 136–37.

80 Aura fell 22 *braccia*, Bourne gives a *braccio* as 0.465 metres. Molly Bourne, *Francesco II Gonzaga: The Soldier-Prince as Patron* (Rome: Bulzoni, 2008), 18.

81 Giovanni Giacomo Calandra to Federico Gonzaga, Mantua, 30 Aug. 1511, AG, b. 2482, c. 115r–116r: 'accadete qui una grande disgratia, che essendo andata la Illustrissima Madama vostra madre a casa di Bagni per visitare la moglie del Conte Bacarino da Canossa paiolata, et volendose partire Sua Excellentia, Aura et la Mamia cagnoline de Sua Signoria se appizzonno insieme per essere stata grande inimicicia tra loro per amore del cane de Alfonso, et ritrovandose su un poggiolo in capo de la scala alto da terra forsi vintidua braza, la povea bella Aura cadde da esso poggiolo su la salicata de la corte, et subito morite con tanto dolore de Madama che non se potria dire. Lo puote ben imaginar ogniuno che sa lo amore che la le portava et quanto meritamente per essere stata la più bella et più piacevole cagnolina che fosse mai. Sua Signoria fu veduta piangere quella sera a

tavola, et mai la non ne parla che non sospira. La Isabella [Lavagnola] piangeva come se le fosse morta sua madre et non se può anchora ben consolare. Non posso già negare che anche io non habbi giettata qualche lachrima. Madama subito fece fare una cassetta de piombo et vi l'ha fatta ponere entro, et credo la tenirà così fin che la se possi mettere in una bella sepoltura [...] Il vostro Zephyro ha perduto una gentil compagna.'

82 Elegies for Aura are collected in a manuscript in the Archivio di Stato in Mantua, Serie Autografi 10, b. 356. See also Walker-Meikle, *Medieval Pets*, 105–107.

83 Signorini, 'A Dog Named Rubino', 318–19. According to Mario Equicola in his *Cronica di Mantua* (Mantua, 1521), Francesco Gonzaga laid his falcons to rest in marble tombs. Bourne, *Francesco II Gonzaga*, 541–42.

84 For instance, Lisa K. Regan, 'Ariosto's Threshold Patron: Isabella d'Este in the *Orlando Furioso*', *MLN*, 120 (2005), 50–69, p. 66, n. 39. These compositions are dismissed as frivolities also by Francesco Malaguzzi Valeri in *La corte di Lodovico il Moro*, i. 585.

85 Luzio and Renier, *La coltura e le relazioni letterarie*, 30–32.

86 Lodovico Gonzaga to Barbara of Brandenburg, August 1467, AG, b. 2099: 'vogliamo gli faciati fare tuti quelli remedii parirano necessarii et non lassarli manchare le suppe grasse né alcuna altra cosa per farlo guarire, s'el sarà possibile. Pur s'el accadesse morire prima che ritorniamo a Mantua, ordinati ch'el sia posto in una cassetta et sotterato lì de drieto al castello dove erano quelle stalle, a ciò che come lui voluntera stasevo in camera nostra cussì nui da la camera possiamo vedere el luogo dove sarà sotterato. Et anche faciamo pensere de farli puoi mettere una petra co l'epitaphio chi ge faremo fare.' Quoted by Signorini, 'A Dog Named Rubino', 317.

87 Signorini, 'A Dog Named Rubino', 318.

88 Such evaluations range from seeing straightforward symbolism related to fidelity or the art of portraiture to more textured readings in which the dog adds qualities of the domestic, intimate, and gentle to the portrait. For instance, Diane H. Bodart, *Tiziano e Federico II Gonzaga: Storia di un rapporto di committenza* (Rome: Bulzoni, 1998); Sheila Hale, *Titian: His Life* (London: Harper, 2012), 262; Charles Hope, *Titian* (London: Chaucer, 2003), 78.

89 Scholars of Titian note the naturalism of the artist's portrayal of dogs. According to Simona Cohen, for example, 'Titian was a brilliant painter of canine species'. *Animals as Disguised Symbols in Renaissance Art* (Leiden: Brill, 2008), 148.

90 Abraham Fleming's translation of Johannes Caius damns lapdogs as 'instrumentes of folly for [women] to play and dally withall [...] these puppies the smaller they be, the more pleasure they provoke, as more meete play fellowes for minsing mistrisses to beare in their bosoms, to keep company withal in their chambers, to succour with sleepe in bed, and nourishe with meate at bourde, to lay in their lappes, and licke their lippes as they ryde in their waggons'. Caius, *Of Englishe Dogges*, trans. Abraham Fleming, 20–21. William Harrison uses this description for his angry condemnation of 'Sybariticall' puppies in Holinshed's *Chronicles of England, Scotland and Ireland* (London, 1587), Volume 1, Book 3, Chapter 7. See Serpell, *In the Company of Animals*, 49–50.

91 Juliana Schiesari, '"Bitches and Queens": Pets and Perversion at the Court of France's Henri III' in Erica Fudge (ed.), *Renaissance Beasts: Of Animals, Humans, and Other Wonderful Creatures* (Chicago: University of Illinois Press, 2004), 37–49.

92 See, for instance, Erica Fudge, 'What Was It like to Be a Cow? History and Animal Studies' in Linda Kalof (ed.), *The Oxford Handbook of Animal Studies* doi:10.1093/oxfordhb/9780199927142.013.28.

93 Haraway, *The Companion Species Manifesto*, 5.

Bibliography

Alekseicheva, Irina, 'Some Statistics on the Breeding of Pallas' Cats in the European Population', Pallas Cat Study and Conservation Program www.savemanul.org/eng/articles/, accessed 26 June 2016.

Atkinson, Niall, 'The Social Life of the Senses: Architecture, Food, and Manners' in Herman Roodenburg (ed.), *Cultural History of the Senses in the Renaissance* (London: Bloomsbury, 2014), 19–42.

Bandello, Matteo, *Novelle* (Lucca, 1554).

Barbieri, Nicoletta Ilaria, 'Cultura letteraria intorno a Federico Gonzaga, primo Duca di Mantova' (tesi di dottorato, Università Cattolica del Sacro Cuore, Milan, 2012).

Barker, Sandra and Aaron R. Wolen, 'The Benefits of Human-Companion Animal Interaction: A Review', *Journal of Veterinary Medical Education*, 35 (2008), 487–95.

Beetz, Andrea et al., 'Psychosocial and Psychophysiological Effects of Human-Animal Interactions: The Possible Role of Oxytocin', *Frontiers in Psychology*, 3 (2012). doi:10.3389/fpsyg.2012.00234.

Belon du Mans, Pierre, *L'histoire de la nature des oyseaux...* (Paris, 1555).

Blaxandall, Michael, *Painting and Experience in Fifteenth-Century Italy* (Oxford: Oxford University Press, 1972).

Bodart, Diane H., *Tiziano e Federico II Gonzaga: Storia di un rapporto di committenza* (Rome: Bulzoni, 1998).

Bourne, Molly, *Francesco II Gonzaga: The Soldier-Prince as Patron* (Rome: Bulzoni, 2008).

Broad, Jacqueline and Karen Green (eds), *Virtue, Liberty and Toleration: Political Ideas of European Women 1400–1800* (Dordrecht: Springer, 2007).

Brown, Clifford M. and Anna Maria Lorenzoni, *Isabella d'Este and Lorenzo da Pavia: Documents for the History of Art and Culture in Renaissance Mantua* (Geneva: Droz, 1982).

Burke, Peter, 'Urban Sensations: Attractive and Repulsive', in Herman Roodenburg (ed.), *Cultural History of the Senses in the Renaissance* (London: Bloomsbury, 2014), 43–60.

Caius, Johannes, *Of Englishe Dogges, the Diversities, the Names, the Natures and the Properties*, tr. Abraham Fleming (London, 1576).

Campbell, Stephen J., *The Cabinet of Eros: Renaissance Mythological Painting and the Studiolo of Isabella d'Este* (New Haven: Yale University Press, 2006).

Castiglione, Baldesar, *The Book of the Courtier*, tr. George Bull (London: Penguin, 1976).

Castiglione, Baldessar, *Il Libro del Cortegiano* (Milan: BUR, 1994).

Chambers, David and Jane Martineau (eds), *Splendours of the Gonzaga* (London: Victoria and Albert Museum, 1982).

Classen, Constance, David Howes, and Anthony Synnott, *Aroma: the Cultural History of Smell* (London: Routledge, 1994).

Clinio Cottafavi, 'Cani e gatti alla corte dei Gonzaga', *Il ceppo quaderno di vita fascista e di cultura* (Mantua, 1934), 8–11.

Cockram, Sarah, 'Interspecies Understanding: Exotic Animals and Their Handlers at the Italian Renaissance Court', *Renaissance Studies* (Special Issue 'The Animal in Renaissance Italy', eds Stephen Bowd and Sarah Cockram), 31 (2017), 277–96. doi:10.1111/rest.12292.

Cockram, Sarah, *Isabella d'Este and Francesco Gonzaga: Power Sharing at the Italian Renaissance Court*, Women and Gender in the Early Modern World (Farnham: Ashgate, 2013).

Cohen, Simona, *Animals as Disguised Symbols in Renaissance Art* (Leiden: Brill, 2008).

Dalton, Heather, 'A Sulphur-Crested Cockatoo in Fifteenth-Century Mantua: Rethinking Symbols of Sanctity and Patterns of Trade', *Renaissance Studies*, 28 (2014), 676–94.

Della Casa, Giovanni, *Galateo overo de' costumi* (Venice, 1558).

d'Este, Isabella, *Selected Letters*, tr. and ed. Deanna Shemek (Toronto and Tempe: Iter Academic Press and Arizona Center for Medieval and Renaissance Studies, 2017).

Dugan, Holly, 'The Senses in Literature: Renaissance Poetry and the Paradox of Perception' in Herman Roodenburg (ed.), *Cultural History of the Senses in the Renaissance* (London: Bloomsbury, 2014), 149–68.

Elias, Norbert, *The Civilising Process*, tr. E. Jephcott (Oxford: Oxford University Press, 1994).

Equicola, Mario, *Cronica di Mantua* (Mantua, 1521).

Equicola, Mario, *De natura d'amore* (Venice, 1525).

Ferrari, Daniela, 'L'inventario delle gioie' in Daniele Bini (ed.), *Isabella d'Este: La primadonna del Rinascimento* (Modena: Il Bulino, 2001), 21–44.

Franchini, Dario, *La scienza a corte. Collezionismo eclettico natura e immagine a Mantova fra Rinascimento e Manierismo* (Rome: Bulzoni, 1979).

Fudge, Erica (ed.), *Renaissance Beasts: Of Animals, Humans, and Other Wonderful Creatures* (Chicago: University of Illinois Press, 2004).

Fudge, Erica, 'Milking Other Men's Beasts', *History and Theory*, 52/4 (2013), 13–28.

Fudge, Erica, 'What Was It Like to Be a Cow? History and Animal Studies' in Linda Kalof (ed.), *The Oxford Handbook of Animal Studies*, doi:10.1093/oxfordhb/9780199927142.013.28.

Guasco, Annibal, *Discourse to Lady Lavinia, His Daughter: Concerning the Manner in Which She Should Conduct Herself When Going to Court as Lady-in-Waiting to the Most Serene Infanta Lady Caterina, Duchess of Savoy*, tr. Peggy Osborn (London: University of Chicago Press, 2003).

Guasco, Annibal, *Ragionamento a D. Lavinia sua figliuola, della maniera del governarsi ella in corte, andando per Dama alla Serenissima Infante D. Caterina, Duchessa di Savoia* (Turin, 1586).

Hale, Sheila, *Titian: His Life* (London: Harper, 2012).

Haraway, Donna, *The Companion Species Manifesto: Dogs, People, and Significant Otherness* (Chicago: Prickly Paradigm Press, 2003).

Haraway, Donna, *When Species Meet* (Minneapolis: University of Minnesota Press, 2008).

Hearne, Vicki, *Adam's Task: Calling Animals by Name* (New York: Knopf, 1982).

Holinshed, Raphael, *Chronicles of England, Scotland and Ireland* (London, 1587).

Hope, Charles, *Titian* (London: Chaucer, 2003).

James, Carolyn, 'Machiavelli in Skirts. Isabella d'Este and Politics', in Jacqueline Broad and Karen Green (eds), *Virtue, Liberty and Toleration: Political Ideas of European Women, 1400–1800* (Dordrecht: Springer, 2007), 57–76.

Jardine, Lisa and Jerry Brotton, *Global Interests: Renaissance Art between East and West* (Ithaca: Cornell University Press, 2000).

Johnson, Geraldine A., 'In the Hand of the Beholder: Isabella d'Este and the Sensory Allure of Sculpture', in Alice Sanger and Siv Tove Kulbrandstad Walker (eds), *Sense and the Senses in Early Modern Art and Culture Practice* (Farnham: Ashgate, 2012), 183–97.

Julius, Henri et al., *Attachment to Pets: An Integrative View of Human-Animal Relationships with Implications for Therapeutic Practice* (Cambridge: Hogrefe, 2012).

Kete, Kathleen, *The Beast in the Boudoir: Petkeeping in Nineteenth-century Paris* (Berkeley and Los Angeles: University of California Press, 1994)

Konecki, Krzysztof, 'Touching and Gesture Exchange as an Element of Emotional Bond Construction. Application of Visual Sociology in the Research on Interaction between Humans and Animals', *Forum Qualitative Sozialforschung / Forum: Qualitative Social Research* 9/3 (2008), www.qualitative-research.net/index.php/fqs/article/view/1154, accessed 23 January 2014.

Loisel, Gustave, *Histoire des ménageries de l'antiquité à nos jours* (Paris: O. Diens, 1912).

Lurati, Patricia, '"To dust the pelisse": The Erotic Side of Fur in Italian Renaissance Art', *Renaissance Studies* (Special Issue 'The Animal in Renaissance Italy', eds Stephen Bowd and Sarah Cockram), 31 (2017), 240–60. doi:10.1111/rest.12293.

Luzio, Alessandro and Rodolfo Renier, 'Il lusso di Isabella d'Este', *Nuova Antologia*, 147 (1896), 441–69; 148 (1896), 294–324; 149 (1896), 261–86, 666–88.

Luzio, Alessandro and Rodolfo Renier, *La coltura e le relazioni letterarie di Isabella d'Este Gonzaga*, ed. Simone Albonico (Milan: Sylvestre Bonnard, 2005).

MacDonogh, Katharine, *Reigning Cats and Dogs: A History of Pets at Court since the Renaissance* (London: St. Martin's Press, 1999).

Malacarne, Giancarlo, *Le cacce del principe: l'ars venandi nella terra dei Gonzaga* (Modena: Il Bulino, 1998).

Malacarne, Giancarlo, *I signori del cielo: La falconeria a Mantova al tempo dei Gonzaga* (Mantua: Artiglio, 2003).

Malaguzzi Valeri, Francesco, *La corte di Lodovico il Moro. La vita privata e l'arte a Milano nella seconda metà del Quattrocento* (Milan: Hoepli, 1913).

McCall, Timothy, 'Brilliant Bodies: Material Culture and the Adornment of Men in North Italy's Quattrocento Courts', *I Tatti Studies in the Italian Renaissance*, 16 (2013), 445–90.

McNicholas, June et al., 'Pet Ownership and Human Health: A Brief Review of Evidence and Issues', *British Medical Journal*, 331 (2005), 1252–54.

Mills, Daniel and Sophie Hall, 'Animal-Assisted Interventions: Making Better Use of the Human-Animal Bond', *Veterinary Record*, 174 (2014), 269–73.

Müller, Corsin A. et al., 'Dogs Can Discriminate Emotional Expressions of Human Faces', *Current Biology*, 25 (2015), 601–5.

Musacchio, Jacqueline, 'Weasels and Pregnancy in Renaissance Italy', *Renaissance Studies*, 15 (2001), 172–87.

Neher, Gabriele and Rupert Shepherd (eds), *Revaluing Renaissance Art* (Aldershot: Ashgate, 2000).

Odendaal, J.S.J and R.A Meintjes, 'Neurophysiological Correlates of Affiliative Behaviour between Humans and Dogs', *The Veterinary Journal*, 165 (2003), 296–301.

Pluskowski, Aleksander (ed.), *Just Skin and Bones? New Perspectives on Human-Animal Relations in the Historic Past* (Oxford: Archaeopress, 2005).

Prizer, William F., 'Una "Virtù Molto Conveniente a Madonne": Isabella d'Este as a Musician', *The Journal of Musicology*, 17 (1999), 10–49.

Raber, Karen, *Animal Bodies, Renaissance Culture* (Philadelphia: University of Pennsylvania Press, 2013).

Regan, Lisa K., 'Ariosto's Threshold Patron: Isabella d'Este in the *Orlando Furioso*', *MLN*, 120 (2005), 50–69.

Richter, Georg Martin, 'The Portrait of Isabella d'Este by Cavazzola', *The Burlington Magazine for Connoisseurs*, 54/311 (1929), 85–92.

Ritvo, Harriet, 'The Emergence of Modern Pet-Keeping', *Anthrozoös*, 1 (1987), 158–65.

Roodenburg, Herman (ed.), *A Cultural History of the Senses in the Renaissance* (London: Bloomsbury, 2014).

Roodenburg, Herman, 'Introduction: Entering the Sensory Worlds of the Renaissance', in Herman Roodenburg (ed.), *A Cultural History of the Senses in the Renaissance* (London: Bloomsbury, 2014), 1–17.

Sanger, Alice E. and Siv Tove Kulbrandstad Walker (eds), *Sense and the Senses in Early Modern Art and Culture Practice* (Farnham: Ashgate, 2012).

Schiesari, Juliana, '"Bitches and Queens": Pets and Perversion at the Court of France's Henri III' in Erica Fudge (ed.), *Renaissance Beasts: Of Animals, Humans, and Other Wonderful Creatures* (Chicago: University of Illinois Press, 2004), 37–49.

Schiesari, Juliana, *Beasts and Beauties: Animals, Gender, and Domestication in the Italian Renaissance* (Toronto: University of Toronto Press, 2010).

Serpell, James, *In the Company of Animals* (2nd ed., Cambridge: Cambridge University Press, 1996).

Sharkin, Bruce S. and Donna Knox, 'Pet Loss: Issues and Implications for the Psychologist', *Professional Psychology: Research and Practice*, 34 (2003), 414–21.

Signorini, Rodolfo, 'A Dog Named Rubino', *Journal of the Warburg and Courtauld Institutes*, 41 (1978), 317–20.

Sykes, Naomi, *Beastly Questions. Animal Answers to Archaeological Issues* (London: Bloomsbury, 2014).

Tague, Ingrid H., *Animal Companions: Pets and Social Change in Eighteenth-Century Britain* (University Park: Penn State University Press, 2015).

Thomas, Richard, 'Perceptions versus Reality: Changing Attitudes towards Pets in Medieval and Post-medieval England', in Aleksander Pluskowski (ed.), *Just Skin and Bones? New Perspectives on Human-Animal Relations in the Historic Past* (Oxford: Archaeopress, 2005), 95–105.

Thornton, Dora, *The Scholar in his Study* (London: Yale University Press, 1997).

Tuan, Yi-Fu, *Dominance and Affection: The Making of Pets* (London: Yale University Press, 2004).

Tuohy, Thomas, *Herculean Ferrara* (Cambridge: Cambridge University Press, 1996).

Walker-Meikle, Kathleen, *Medieval Pets* (Woodbridge: Boydell Press, 2012).

Welch, Evelyn, 'New, Old and Second-hand Culture. The Case of the Renaissance Sleeve', in Gabriele Neher and Rupert Shepherd (eds), *Revaluing Renaissance Art* (Aldershot: Ashgate, 2000), 101–19.

Wilson, Elizabeth and Wilson Stephens, *Four Centuries of Dog Collars at Leeds Castle* (London: Leeds Castle Foundation, 1979).

3   Equine empathies

Giving voice to horses in early modern Germany

Pia Cuneo

Empathy, animal autobiography, and anthropomorphism

Any survey of early modern imagery reveals a colourful parade of horses in virtually all media. Albrecht Dürer's well-known engraving *Knight, Death and the Devil* (1513) provides a typical example of just such imagery (Figure 3.1).[1] Most early modern images, like Dürer's print, depict the horse as imposingly powerful but utterly obedient to the rider's will; horse and human exist together in a state of perfect harmony, without any hint of conflict or discord.[2] In the engraving, the horse trots energetically but calmly forward while the rider easily, even casually, directs the movement with only one hand holding the double reins. Following philosophical and artistic traditions stretching back to classical antiquity, early modern representation of the horse's obedient submission is understood both to communicate and to prove the genuine authority of his rider. The horse functions as a cultural symbol for virtuous human mastery over a number of dangerously capricious phenomena, including emotions, political states, and nature itself.

While strategies of visual representation at play in images like Dürer's print encode a host of social, political and cultural ideologies, they also render the horse as a living, sentient, particular being practically invisible. Animal studies challenges historians of every discipline to search beyond representation in order to get as close as possible to the lives, experiences and even agency of animals.[3] While this enterprise is fraught with difficulties ranging from the epistemological to the methodological, intellectual and ethical imperatives compel us to think about history and culture in ways that are sensitive to the interconnectedness between – some would even say the consubstantiality of – humans and animals, both in the past and in the present.[4]

Viewing early modern equestrian imagery through an animal studies lens changes the kinds of questions we ask of the works. It is no longer enough to ascertain what gait or movement the horse is performing in the image, what virtue/vice the animal symbolizes, or even what ideologies and practices inform the horse's representation. Now the questions include: what was it like for horses to interact with their riders? What did they experience in being trained to perform movements they are represented in visual imagery

Figure 3.1 Albrecht Dürer, *Knight, Death, and the Devil* (engraving, 1513), © The Trustees of the British Museum.

as executing? What were the possibilities for equine agency in a horse's interaction with a human?

My research on early modern German hippology has yielded three sources that provide some answers to these questions. In this essay, I use these texts as suggestive sources that provide an insight into horses' lives in the early modern period. Written in the mid-sixteenth and early seventeenth centuries, these three texts attempt to provide a horse's point of view of human-animal interaction by giving them voices with which to speak. Written of course by humans, the texts nonetheless endeavour to articulate what horses may have experienced and how they may have responded. At the bases of these texts lie observations about contemporaneous horse-human interactions that reveal close attention to the equine side of the equation.

I have designated these texts as empathetic, rather than sympathetic, because of the deeper level of engagement involved in empathy. To feel sympathy with fellow creatures is to be concerned about them and to feel sorry for their misfortune; to experience empathy moves beyond merely noting another's problems with concern to actually sharing the other's feelings, to experiencing those feelings yourself.[5] In order to write about a horse's feelings – to give words to an animal's experiences – the three German authors had to share what they perceived to be those feelings and experiences, and thus their writing was a deeply empathetic act. Furthermore, as we will see, the goal of their texts is not simply to draw readers' attention to the suffering experienced by horses at the hands of humans, but to awaken the readers' empathy and create compassion in order to effect change.

Historical, neurological factors and historically specific cultural practices support my argument that contemporaneous readers responded empathetically to the three German texts. The art historian David Freedberg has drawn on neurological research in order to understand viewers' somatic and emotional responses to visual imagery. He cites scientific studies that show how the same responses are triggered in the nervous system of the viewer of an image as would have been triggered in a person who was actually engaged in the movement/experience depicted in an image.[6] Other studies he references link a viewer's emotional/empathetic response to the emotional response of another person observed by the viewer.[7] Thus there is a neurological basis for the experience of empathy based on visual perception. We are, it seems, hard-wired for empathy and have been for millennia.[8] Although the sources interrogated here are not visual images but rather texts, their insistent and detailed descriptions enable and encourage a reader to vividly picture in her/his imagination the texts' accounts. These compelling mental images of another's suffering would provide the basis for the readers' empathetic response.

Not only would the activation of specific neurons and parts of the brain account for such a response, I speculate that early modern cultural practices also shaped the reception of these texts. Their authors, Hans Sachs, Christoph Lieb, and Gabriel von Danup, include uncomfortably graphic descriptions of horses' maltreatment by humans. In these descriptions, the horses sweat and bleed profusely, they are whipped and spurred, they are shouted at and cursed, their bodies are contorted and broken by bearing heavy loads, and they die miserable deaths. Early modern readers would have been familiar with a similar catalogue of abuse and suffering. In order to facilitate and stimulate personal piety, sermons and devotional treatises often focused on the Passion of Jesus, describing in lurid detail Jesus' physical and emotional torment.[9] Bodily fluids such as sweat, blood, and spittle feature prominently in these accounts. They relate how Jesus was whipped and his skin pierced by the crown of thorns, how his tormenters shouted and cursed at him, how his body was broken down further under the weight of the cross he was forced to drag to Golgotha, and how he died a miserable death nailed to the cross. All of these descriptions are accompanied by exhortations that the listeners/ readers should immerse themselves deeply in these narrated events to the point that they are able to experience what Jesus did, to feel the same emotional anguish and physical suffering. The goal of this empathetic response was to effect change: for the believer to understand that her/his sins were the ultimate source of Jesus' sacrificial death – and its torturous preamble – and thus to repent.

As Susan Karant-Nunn has argued, this treatment of the Passion narrative is more characteristic of Christianity prior to the Reformation.[10] She demonstrates that Martin Luther (1483–1546) rejected the rhetorical and contemplative strategy of providing grisly accounts of the Passion intended to bring their listeners/readers to emotional and physical outbursts such as weeping and wailing. Karant-Nunn argues that Luther emphasized the emotions of consolation and love as appropriate responses for a believer's contemplation of the Passion: 'Luther did not want the laity to figuratively flagellate themselves and to weep. Having mastered the story, they were to move quickly beyond it to God's love for them, and Christ's obedience in fulfilling the ordained atonement. On this they were to rest their thought and their feelings'.[11] Nonetheless, Karant-Nunn also notes that how subsequent Lutheran and reformed theologians treated the Passion in their texts and sermons varied considerably and that some continued to dwell on painful details in order to encourage the believer to seek atonement.

The texts by Sachs, Lieb, and von Danup were all produced well after the process of reformation had begun in the German territories, but Karant-Nunn's qualifications suggest that even readers who came from Lutheran or other reformed traditions may still have been familiar with engaging empathetically with a viscerally evocative narrative, like the Passion, and would have brought that familiar practice of engagement to the texts on horses.

This entire process of empathetic response on the parts of the authors Sachs, Lieb, and von Danup and of their readers, as well as the ideally ensuing changes of attitude and behaviour, must have been rooted in the observable behaviour of horses. They communicate their experiences through their bodies so vividly that the authors feel enabled and compelled literally to give the horses voices with which to speak. Thus we may think about horses exercising a kind of agency, albeit in cooperation with humans, to effect potential change.

Giving voice to horses situates the texts by Sachs, Lieb, and von Danup within a literary genre that we would call today 'animal autobiography.' All three German texts are autobiographical because in each the horses appear to relate their own life experiences. For Anglophones, the most famous example of 'animal autobiography' may be Anna Sewell's *Black Beauty* (1877), but this is actually a venerable genre rooted in classical antiquity.[12] For example, in an epigram attributed to the Greek poet Aulus Lucinius Archias (fl. 120–60 BCE), a horse laments that despite the many brilliant victories he claimed in his youth at the Panhellenic games, now as an old horse he provides power for a grindstone by walking around in endless circles, goaded into ceaseless motion by frequent lashings of the whip.[13]

Writers of 'animal autobiography' provide easy targets for various critiques that range from sentimentality and triviality, to outright ventriloquism and anthropomorphism. But a number of scholars defend the genre. Like images, and like any other kinds of texts, these 'autobiographies' must also be understood as representations. As Karla Armbruster notes, refusing the possibility that humans could ever speak for animals based on 'the acceptance of such a radical gap between species simply perpetuates anthropocentric assumptions about human uniqueness and superiority.'[14] For Ryan Hediger, the main issue involved in writing and reading 'animal autobiography' is not whether or not human language can accurately reveal an animal's mind. He advocates thinking about language as a tool rather than as a vehicle for epistemological certainties. Hediger labels this kind of language as 'rhetorical', and he claims that it allows animal autobiographies to function in a potentially useful way for humans and animals in the real world. For Hediger, the main issues are:

> How can texts representing nonhuman animals be useful to us and to animals? How can they help us, however imperfectly, to understand other lives? So, rather than foreswearing the attempt to think through nonhuman lives, often in fear of the dreaded specter of 'anthropomorphism,' a rhetorical approach allows that all language use is finally situated, pragmatic, not universal.[15]

Language does not offer a perfect system granting us unimpeded access to an animal's experiences. In addition, it is difficult to imagine that our understanding of animals could ever completely transcend the bounds of anthropomorphism. But just because we cannot get it completely right does not mean that our attempts to think about and discuss the lives of animals have no value whatsoever. In fact, Erica Fudge points to a potentially ethical dimension of anthropomorphism evident in texts like 'animal autobiographies':

> We may regard the humanization of animals that takes place in many narratives as sentimental, but without it the only relation we can have with animals is a very distant, and perhaps mechanistic one. As well as this, anthropomorphism might actually serve an ethical function [...]. By gaining access to the world of animals, these books offer a way of thinking about human–animal relations more generally, and potentially more positively.[16]

In turning now to the texts by Sachs, Lieb, and von Danup, I take my cue from Hediger and Fudge in looking beyond the obvious charge of anthropomorphism that could be lodged against all three. Instead, I will think of these texts as pragmatic and 'useful [...] to animals' (Hediger), and as ethical narratives intended to encourage early modern readers to think 'about human-animal relations [...] potentially more positively' (Fudge). My point of departure shall be that the intended function of these three texts – at least in part – was to improve the lives of horses. In order to fulfil that function, the authors describe some of the practices and conditions to which horses were subjected. In so doing, they offer the historian vital information about early modern horse-human interactions that moves well beyond all the pretty pictures of the masterly rider on his happily obedient mount.

Hans Sachs (*c*.1557), Christoph Lieb (1616) and Gabriel von Danup (1623)

European texts on horsemanship, including those written in German, unfailingly include passages that eloquently praise horses for their beauty, nobility and utility. For example, in the introduction to a book on the treatment of equine ailments printed in 1583 in Strasbourg, the horse is lauded for his special role in the display of noble identity: 'With his exceptionally beautiful form and figure, his brave, aristocratic disposition, and his proud, erect body, the horse is particularly well-suited to highlight the honour and preeminence of the nobility.'[17] Certainly passages such as these were helpful in justifying the production and encouraging the consumption of such books, but they also must have spoken to an appreciation for horses and the fundamental roles they played in many people's lives. Nonetheless, admiration and appreciation constitute only a small portion of attitudes about horses in early modern Germany.

Hans Sachs (1494–1576), the famous shoe-maker, *Meistersinger* and prolific poet from Nuremberg, reveals another subset of attitudes and their attendant practices in a prose-poem written and published *c*.1557. According to these attitudes, the horse was simply an object to be used for a variety of human needs. These uses did not in any way oblige humans to be concerned with or attend to the animal's wellbeing. Consequently, the life of a horse as described in Sachs' text was miserable from start to finish. Sachs wrote on a strikingly wide variety of topics, including texts favourable to Martin Luther and the process of reformation that his native Nuremberg had officially embraced by 1525.[18] Animals appear as characters in many of the approximately 1,700 prose-poems that he composed over his lifetime. However, only the poem Sachs wrote in *c*.1557 features a horse who tells his entire life-story from his own perspective.[19]

Sachs' poem, entitled *The Miserable, Lamenting Horse-Hide* (Ger. *Das ellend klagend Rosshaut*), opens on a Monday morning.[20] A badly hungover cobbler enters his workshop and gets out a piece of horse-hide leather with the intent to make shoes. But when he begins to cut the leather with his shears the horse-hide exclaims: 'Oh! Stop cutting into me! How much more misery must I endure?' The greatly astonished cobbler asks the horse-hide who he is, and the piece of leather responds: 'I am the old skin of a horse and I suffered much during my life.'[21] Thereupon follows the hide's lengthy and detailed narrative of his utterly lamentable life.

The horse is born on a farm but essentially neglected. He is bought by a profit-hungry horse-trader and sold to an impoverished nobleman who supports himself by highway robbery. This is a very strenuous life and because the nobleman is poor, he does not give the horse much to eat. When this noble highwayman is arrested, the horse is sold to a wealthy burgher who feeds him well enough but rides him hard and rides him badly. During one of the many tournaments in which the pretentious burgher participates, the horse is injured and becomes lame. As a result, he is sold to a wagoner who hitches the horse to stupendously heavy loads that the animal must haul up and down mountains, through snow, rain, and mud and in great heat and cold. The wagoner regularly beats him and gives him poor fodder. After ten years of such toil, the horse is so broken down that he can no longer pull large wagons so he is sold to pull small carts. He is so feeble by this point, however, that it is not long before, as a wretched bag of skin and bones, with hanging head and drooping ears, he is brought to the skinner and slaughtered. Even as a carcass the horse suffers, being gnawed upon by the skinner's dog.

Moved by the empathy he experiences by listening to the horse-hide's story, and by the hide's plea to spare him, the poem's cobbler changes his course of action.[22] He explains to the hide that he is obligated to make shoes, and that the hide is the only piece of leather he has in his workshop. Nonetheless, the cobbler fulfils the hide's wish: instead of using the horse's skin to make farmers' boots to be worn outside in filth and slop, the cobbler makes dainty slippers to be worn inside by ladies. While it is true that the cobbler's change of course in the poem does not do much for the way living horses are treated by humans, he shows the hide what mercy he can. It is up to the reader, who has followed the story and who presumably will not be dealing with talking horse-hides but with the animals themselves, to perform what mercies s/he can.

The horrendous treatment of the horse meted out by the men he encounters certainly serves to highlight human thoughtlessness and selfish cruelty. The over-arching theme of Sachs' prodigious oeuvre is the critique of human behaviour with the intent of inspiring moral improvement. Sachs' texts are clearly didactic in nature. They deal with quotidian subject matter and use easily understood, even colloquial, language in order to prove their points. The readers of Sachs' *The Miserable, Lamenting Horse-Hide* would most likely have come from a broad spectrum of literate and semi-literate urban dwellers, both in Nuremberg and beyond. Thus the potential audience of the text was both large and varied.[23] The apparent goal of Sachs' text is to encourage this audience to recognize some human behaviour as cruel and wrong and to empathise with the horse. For that to happen, Sachs' narrative of what the horse has experienced must have been recognisably familiar to the audience. Thus, despite the poem's additional functions such as praising the cobbler's craft (a description of which takes up just over half of the poem) and criti-quing human failings, the text in fact can tell us something about what life was presumably like for some horses. The text both springs from and strives to incite an empathetic response that will result in an awareness of the plight of these animals and a change of behaviour.

At this point, we should ask how the cobbler and poet Sachs would have been familiar with horses' experiences and what could have informed his empathetic response to them. Answers to these questions remain speculative. As a cobbler, Sachs would perhaps have had dealings with the local skinner/ knacker (*Schinder/Abdecker*). Because their meat was generally not for human consumption, horses at the end of their lives were not given to a butcher (*Metzger*) for slaughter, but instead, as in Sachs' poem, handed over to the skinner or knacker, who would skin the carcass and either return the hide to the owner for a fee or retain it to sell on to merchants or to craftsmen who worked with leather, such as tanners and saddlers, and also perhaps to cob-blers.[24] In Sachs' poem, the horse-hide is sold by the skinner to a merchant, who in turn sells it to a tanner, who sells it to the poem's cobbler. Perhaps the involvement of several different middlemen described by Sachs for the pro-curement of the hide was typical for production in a large city like Nurem-berg, as opposed to smaller towns and villages, where such middlemen would presumably not be needed. But their intervening presence also serves to inti-mate a desired if not actual distance between the cobbler – the representative of Sachs' own honourable craft – and the skinner, often a poor, uneducated man whose duties also included other useful but unsavoury activities such as slaying stray dogs, cleaning out latrines, and assisting the executioner.[25]

Sachs certainly would have seen horses in the streets of Nuremberg pull-ing carts and wagons and being ridden by other people; presumably at some time he himself would have travelled somewhere on horseback. The same was probably true for the many readers of Sachs' text as well. That horses even occasionally wandered loose in the streets is indicated by the existence of a civil servant (the *Löw*) in Nuremberg whose job it was to round up stray horses and pigs and to return them to their owners once a fine had been paid.[26] That Sachs may himself have owned horses is suggested by the fact that his personal library included a book on how to cure equine ailments.[27] In these ways, Sachs would have gained familiarity with a variety of horse-human interactions.

The heartlessly instrumentalist attitudes towards horses that Sachs excori-ates in his poem may have been part of a general attitude about animals indi-rectly informed by the writings and ideas of Thomas Aquinas (*c*.1225–1274) that dominated the later Middle Ages and continued to enjoy authority in the early modern period.[28] Aquinas' standpoint may be summed up by the following quotation: 'for by divine providence, they [animals] are intended for man's use according to the order of nature. Hence it is not wrong for man to make use of them, either by killing or in any other way whatsoever.'[29]

Aquinas does concede that humans should avoid treating animals cruelly. But the issue for Aquinas is not the suffering the individual animal would experience under those conditions but that cruelty to animals leads to cruelty to other humans. Sachs' poem parts company with Thomist attitudes since nowhere in the horse-hide's lamentable list of the abominations he experiences is any concern voiced that such despicable behaviour would transfer over into inter-human relations. It is unclear why Sachs appears to reject a constellation of attitudes that held sway for centuries. It is tempting to surmise that Martin Luther's views may have informed Sachs' standpoint because Sachs publicly responded so positively to Luther and his ideas. But work has yet to be done on Luther's perspective on animals. Preliminary research suggests the reformer's basically instrumentalized view of nature while also pointing to possibilities for his more open and direct engagement with animals.[30]

Evidently, Sachs knew what he was talking about because the maltreatment of horses described in his poem is also described in a horsemanship manual written by Christoph Jakob Lieb. What we know about Lieb is only what we can glean from the book's introductory text in which Lieb identifies himself as having served as a *Bereutter* for the Lutheran Electoral Prince of Saxony, Christian II (1583–1611) at his court in Dresden.[31] *Bereutter* were men employed to exercise and train the horses belonging to and stabled at the court.[32] The horse in Sachs' poem reports that his first owner, the noble highwayman, regularly 'hammered [my sides] with his sharp spurs and beat me hard around the ears.'[33] In his manual *The Practice and Art of Riding* (Orig. *Practica et Arte di Cavalleria. Übung und Kunst des Reitens*), published in 1616, Lieb criticizes just such practices, among others:

> Punishment of a horse who will not go forward may be given in a number of ways, depending on the individual animal [...]. But I do not like the punishments that some people use, such as jabbing the horse with spikes, or burning him with fire, or wrapping a rope around his genitals and then yanking on it when he attempts to disobey by standing still. [...] Others are accustomed to punishing the horse by beating him with whips or clubs between his ears and on his head; I do not like this either because two plates of the horse's skull join together at this place and by striking the horse here one can easily break the skull apart and if that happens the horse will die.[34]

Lieb's manual and Sachs' poem also intersect on the topic of bits. Sachs' horse describes how his second owner, the pretentious burgher, 'tortured me first with one kind of bit, and then with another, until I didn't know what to do, whereupon he beat and whipped me harshly.'[35] The illustrated title-page to Lieb's riding manual takes up a similar point, and this title-page serves as my second example of an empathetic, 'animal autobiographical' text. On this page, in addition to the full title of the manual, we find a brief poem that is written as if by the horse himself (Figure 3.2).

Figure 3.2 Title-page engraving to Christoph Jakob Lieb, *Practica et Arte di Cavalleria* (Dresden: Gimel Bergen, 1616). © Herzog August Bibliothek Wolfenbüttel <http://diglib.hab.de/drucke/3-2-bell-2f-2s/start.htm,%20image%2000001>.

In this poem, the horse refers to the many inept riders to whom he is subjected. Although they are the ones making the mistakes and riding poorly, they blame the horse. To fix the problems they themselves cause, the riders experiment with all kinds of different bits in the horse's mouth. Lieb's horse laments that even though he is very willing and tries to accommodate these riders, what saddens him most is that his only reward from them is their ungratefulness. In the manual, Lieb discusses horses that are 'hard in the mouth' (*hartmeulig*). He argues passionately against using harsh bits to solve a problem that often has nothing to do with the horse's mouth but actually has its origins elsewhere, such as in feet that are sore or backs that are weak. He admonishes his readers:

> Just remember that the horse is made of flesh and blood [...]. If you attack and molest a horse with harsh bits, if you rip open and wound his mouth, which is what these kinds of bits do, not only are you not going to solve the problem, you are going to ruin the horse.[36]

As the above quotation intimates, Lieb writes his manual as a horse-trainer for other horse-trainers. He specifies this readership in the first part of his lengthy and descriptive title: *The Practice and Art of Riding in which the Horse-trainer [...] Should be Experienced and Practiced [...].* (Ger. *Übung und kunst des Reitens/ in welcher der Bereuter / [...] erfahren und geuebt sein sol [...]*). The last thing a horse-trainer would want to do is to ruin an expensive animal owned by a noble client, and Lieb's text explains how to avoid professional disaster. His text, which presents horses as fellow creatures made of flesh and blood and who possess human emotions like sadness is surely meant to influence the attitudes and actions of trainers by encouraging them to abandon or eschew cruel and violent practices and to work with their charges in a manner appropriate to an individual horse's physical and emotional constitution.

Further along in Lieb's interminable title he specifies a further kind of readership for his text. The title states that Lieb has written his book *to the especial honour and for the pleasure of all those who love and are partial to this noble and knightly art* [of riding] (Ger. *Allen Liebhabern / und dieser Adelichen Ritterlichen Kunst zugethanen / zu sonderbahren Ehren und gefallen*). Here we are probably dealing with an audience not of horse-trainers but the men who employ them, members of the higher nobility such as Christian II's successor Johann Georg (1585–1656), to whom Lieb's book is formally dedicated. The book could be used by these noble employers as a reference to ascertain whether or not their horse-trainers are following correct procedures and implementing appropriate practices. The book certainly serves to signal Lieb's own knowledge as well as his attitudes about horses – attitudes that emphatically embrace the empathetic. He may well have been interested in communicating these things to Johann Georg, the brother of Lieb's former employer, Christian II, who had died five years prior to the book's publication. Lieb may have intended the text to draw Johann Georg's attention to his dead brother's faithful servant and skilled horseman in the hope of finding subsequent employment at the Dresden court. In light of this possibility, we can speculate that Lieb's empathy with the horses he trains, made so evident in his book and even strikingly advertised on its title-page, may have been regarded by him as an attribute that would aid him in acquiring noble patronage. This is not to imply that Lieb's empathy was necessarily disingenuous, but to suggest that empathy was valued as a professional qualification in certain circles, including at the Saxon court in Dresden.

The third 'animal autobiography' was also written by a former court horse-trainer. In his horsemanship tract published in 1624, Gabriel von Danup tells the reader that he has recently retired from his post at the Electoral Prince of Brandenburg's court and that although he is 51 years old, he can easily ride and train five or more horses a day.[37] This information seems to indicate that he is currently seeking employment. But it is his remarkable pamphlet, published a year earlier (in 1623), the lengthy title of which is most aptly abridged to *The Horses' Supplication*, that contains an arresting and powerful example of equine empathy.[38] The pamphlet describes a fantastic dialogue between Wilhelm Ludwig, Count of Nassau, renowned in his day as a consummate horseman, and the Italian riding master Pirr'Antonio Ferraro of Naples.[39] At the time the pamphlet was published, both men were already deceased. Their dialogue, which is actually a pointed debate, takes place on Mount Parnassus before the god Apollo. The ostensible reason for the debate is that Count Wilhelm has been asked by horses to bring to Apollo their petition for protection against tyrannical riders. Accordingly, Apollo has granted Wilhelm an audience during which the count reads the petition aloud.

The horses' petition is heart-breaking.[40] In the text, they describe how their riders train them. The description features many of the practices we have already encountered and adds several more. The horses' talk moves from being whipped and beaten about the head and between the ears, to riders using their spurs against the horses' sides as hard as they possibly can, to being stabbed from the ground by unmounted men with iron spikes and to having their mouths ripped open and bloodied. In addition, the horses describe harsh use of the caveson-rein, jerked with such force and frequency across the thin skin of their noses that the cartilage is laid bare; the practice of tying their heads down to their chests or to one side or another and leaving them thus constrained and contorted for hours on end; and being constantly shouted at and cursed.

As a trainer of horses, von Danup, like Lieb, was certainly an eyewitness to such brutal practices. Von Danup's desire and ability to describe those practices from the perspective of the horse strongly suggests that he experienced a deep empathy with the animals that he hoped to awaken also in his readers. He certainly wanted to effect change and improve the horses' lives. The horses' supplication to Apollo concludes with their plea that the god should command riders to stop tormenting them with their ignorantly barbaric methods and instead to train them according to true skill and science

(*Wissenschaft*). Count Wilhelm suggests that professional horsemen should work together towards this goal, for the wellbeing of the horses and for the benefit of their riders. Apollo agrees and the text concludes.

The pamphlet is not dedicated to anyone and there are no introductory remarks. However, the intensely detailed discussion of riding techniques, the identity of the clear protagonist as the Count of Nassau, and the conceit of having this debate take place on Mount Parnassus indicates that the intended readership must have included horsemen from elite circles. This was the audience von Danup hoped to impress, perhaps to gain patronage and support or simply to enhance his reputation. Von Danup's passionate empathy seems to spring from profound knowledge, superior skill, and a keen sense of mercy. As in Lieb's case, these qualities may well have been genuine while they also may have enhanced von Danup's professional prospects.

The virtue of empathy also allows von Danup to criticize the Italians, hailed by many contemporaries as preeminent in the art of riding.[41] In sharp distinction to von Danup's protagonist, Count Wilhelm, the Neapolitan riding master Pirr'Antonio rejects the horses' supplication as completely baseless. According to Pirr'Antonio, horses are essentially uncooperative creatures who, with their innate wrathfulness (*Zorn*) and stubbornness (*Halßstarrigkeit*), resist the riders' virtuous efforts to train them; thus the riders are compelled to punish the animals' obstinacy.[42] During the course of their debate, Pirr'Antonio's methods and those of the entire school of Italian riding that he represents are unmasked as deeply flawed. Von Danup wishes to persuade his readers that the Italians ride poorly. He describes how they focus mainly on forcing the horse's head down and driving the animal's weight onto his forehand. This is not how horses are naturally meant to move. Weighting the forehand means that the horse is out of balance and the weaker parts of his body are forced to do the work and carry the weight, which means that the horse is uncomfortable and will also eventually break down. But because they do not know any better, and because they have no empathy with the horses, the Italians persist in their malpractices and force the animals into submission by whipping, spurring, and beating them. The correct way to ride, von Danup maintains, is to do the exact opposite. The horse must be encouraged to carry his weight on his hindquarters where he is most strong. In this way, the horse is able to balance himself and move freely and elegantly, just as nature intended. Because the Italians lack knowledge, skill, and empathy, they work their horses in a way that goes against the animals' physical and emotional natures. As a result, their horses are resistant and obstreperous. Empathy thus functions as the fulcrum of an argument intended to discredit the rival Italian school of riding and to elevate northern methods. By knowledgeably criticizing the purported masters of riding and pointing out the error of their ways, von Danup could clearly demonstrate the depth of his own professional skill and experience. As a German horseman, his acute analysis of Italian practices also signals a legitimate transalpine contribution to the pan-European world of horsemanship.

At the end

Sachs, Lieb, and von Danup appear to respond empathetically to their observed behaviour of horses by writing texts that ideally 'would be useful [...]
to animals' (Hediger) by encouraging early modern readers to think 'about
human–animal relations [...] potentially more positively,' (Fudge) that is,
from the horse's perspective. Recognizing that it was presumably real horses
that set this process in motion is also to recognize that they exercised a degree
of agency. This is not to say that empathy was the authors' sole motivation:
moral critique (Sachs), self-advancement (Lieb), even regional chauvinism
(von Danup) had their roles to play. Nonetheless, the efforts of these authors
to construct and explore an equine viewpoint provide striking evidence of an
emotion that the authors may well have genuinely experienced.

According to these texts, horse-human interactions in the early modern
period included horrible cruelty. Horses' genitals were rubbed raw with
ropes, their heads were beaten in with clubs, and their mouths were ripped
open by metal spikes. Sachs' horse says that he experienced suffering, misery
and torture and even describes himself as a martyr. In the horses' supplication
written by von Danup, the horses beg repeatedly for mercy at the hands of
their riders and for protection against the senseless violence perpetrated upon
them. They describe their lives as a tragedy of toil, and like Sachs' horse,
as martyrdom. To return to the questions posed at the beginning of this
chapter, these experiences of the horse and his interaction with the humans
who trained him as described in the texts are rendered completely invisible
by early modern equestrian imagery that instead presents a thoroughly anthropocentric fantasy of effortless human control over willingly subservient
nature.

There is also no place in this fantasy for the way in which these horses
lived or the way in which they died. The end of a horse's life was often
lamentable.[43] Both Sachs' and von Danup's horses speak about what happens
once they have been injured and lamed by their abusive riders. The horses are
sold to carters and wagoners to be used primarily for hauling until they are
unable to work. At this stage, they are handed over to the knacker who kills
and skins them. The final resting place of these once-magnificent creatures
is either a shallow grave in the earth or on a rotting pile of carrion in the
knacker's yard.

Archaeological investigations of a yard in Emmenbrücke (in the Swiss
Canton of Zürich) revealed skeletons of over 700 animals, 379 of which were
horses.[44] The yard was in use from 1562 to 1866 and the individual skeletons
have not been dated. Nonetheless, the zoo-archaeological findings are suggestive. The largest number of dead horses were between the ages of 5–12 years
(189, compared to 29 horses under the age of 4 and 23 over the age of 20).[45]
The archaeologists examined 100 bone-samples from the equine skeletons.
Seventy-three out of 100 were characterised by pathological changes in the
spinal column, hock and toe joints, stemming either from degenerative joint

disorders or mechanical trauma.[46] Both of these sources indicate an exploitation of the horse beyond what the animal could sustain. The archaeologists conclude 'that the horses used for riding and hauling were exposed to particular strains and that they were completely used up before they found their final resting place in the knacker's yard.'[47] That many of these horses were 'completely used up' between the relatively young ages of 5 and 12 indicates how full of intense toil their brief lives had been. The archaeologists also reveal that the knacker's yard was located on the same grounds where criminals were executed, evidently not an unusual combination of activities within one locale.[48] Thus it must have been a site of special horror and suffering for both horses and humans. These archaeological investigations strongly suggest that the animal autobiographies considered in this essay are more than mere works of fiction. Instead, they appear to be firmly grounded in the brutal realities manifest in zoo-archeological evidence.

Notes

1 See Pia F. Cuneo, 'The Artist, His Horse, a Print, and its Audience: Producing and Viewing the Ideal in Dürer's *Knight, Death, and the Devil* (1513)', in Larry Silver and Jeffrey Chipps Smith (eds), *The Essential Dürer* (Philadelphia: University of Pennsylvania Press, 2010), 115–29.

2 One rare exception to early modern representations of the horse as controlled and obedient are the works of the German painter and printmaker Hans Baldung Grien (*c.*1484–1545). See Pia F. Cuneo, 'Horses as Love-Objects: Shaping Social and Moral Identities in Hans Baldung Grien's *Bewitched Groom* (ca. 1544) and in Sixteenth-Century Hippology', in id. (ed.), *Animals and Early Modern Identity* (Farnham: Ashgate, 2014), 151–68.

3 The now-classical articulation of this challenge is Erica Fudge, 'A Left-Handed Blow: Writing the History of Animals', in Nigel Rothfels (ed.), *Representing Animals* (Bloomington: Indiana University Press, 2002), 3–18.

4 Critical engagements with the work of Emmanuel Levinas have recently interrogated the relationships among animal ethics, philosophy and literature. See Matthew Calarco, *Zoographies: The Question of the Animal from Heidegger to Derrida* (New York: Columbia University Press, 2008), 55–77.

5 Oxford English Dictionary Online, Oxford University Press, www.oed.com, accessed 14 Sep. 2016, s.v. 'sympathy, n.' (sense 3.c) cf. 'empathy, n.' (sense 2.a).

6 David Freedberg, 'Memory in Art: History and the Neuroscience of Response', in Suzanne Nalbantian, Paul M. Matthews, and James L. McClelland (eds), *The Memory Process: Neuroscientific and Humanist Perspectives* (Cambridge: MIT Press, 2010), 341.

7 David Freedberg, 'Empathy, Motion and Emotion', in Klaus Herding and Antje Krause Wahl (eds), *Wie sich Gefühle Ausdruck verschaffen: Emotionen in Nahsicht* (Berlin: Driesen, 2007), 41.

8 According to the work of primatologist Frans de Waal, this is another thing we share with animals. Frans de Waal, *The Age of Empathy* (New York: Three Rivers Press, 2009).

9 Susan Karant-Nunn, *The Reformation of Feeling: Shaping the Religious Emotions in Early Modern Germany* (Oxford: Oxford University Press, 2010), 15–62.

10 Karant-Nunn, *Reformation of Feeling*, 63–99.

11 Karant-Nunn, *Reformation of Feeling*, 97.

12 There are many valuable and critical analyses of Sewell's work. For an excellent discussion of the book's eighteenth- and nineteenth-century contexts, see Diana Donald, *Picturing Animals in Britain 1750–1850* (New Haven: Yale University Press, 2007), 198–232.

13 Cited in Wolfram Martini, Jochen Küppers and Manfred Landfester, 'Römische Antike', in Peter Dinzelbacher (ed.), *Mensch und Tier in der Geschichte Europas* (Stuttgart: Alfred Kröner Verlag, 2000), 98.

14 Karla Armbruster, 'What Do We Want from Talking Animals? Reflections on Literary Representations of Animal Voices and Minds', in Margo DeMello (ed.), *Speaking for Animals* (New York: Routledge, 2013), 21.

15 Ryan Hediger, 'Our Animals, Ourselves. Representing Animal Minds in *Timothy* and *The White Bone*', in Margo De Mello (ed.), *Speaking for Animals* (New York, Routledge, 2012), 35.

16 Erica Fudge, *Animal* (London: Reaktion, 2002), 76–77.

17 *Ein Newe und bewerte Rossartzney* (Strasbourg, 1583), sig. Aii^{r-v}. All translations from the German are my own.

18 Hans Joachim Behr, 'Hans Sachs: Handwerker, Dichter, Stadtbürger. Versuch einer Würdigung anläßlich der 500. Wiederkehr seines Geburtstages', in Dieter Merzbacher (ed.), *500 Jahre Hans Sachs. Handwerker, Dichter, Stadtbürger* (Wiesbaden: Harrossowitz, 1994), 9–16. See also Gerald Strauss, *Nuremberg in the Sixteenth Century* (Bloomington: Indiana University Press, 1976), 154–86.

19 In many of Sachs' works, the animal characters speak but the texts in which these characters are given voice are fables. See Winfried Theiß, *Exemplarische Allegorik. Untersuchungen zu einem literarischen Phänomen bei Hans Sachs* (Munich: Wilhelm Fink Verlag, 1968), 122–36.

20 Hans Sachs, *Die ellend klagend Rosshaut* (Nuremberg, ca. 1557), reproduced in id., *Werke*, eds Adelbert von Keller and E. Goethe, 26 vols (Tübingen: H. Laupp, 1870–1908), v. 146–53. See also Renate Freitag-Stadler (ed.), *Die Welt von Hans Sachs: 400 Holzschnitte des 16. Jahrhunderts* (Nuremberg: Hans Carl, 1976), cat. no. 299 on 259, illustration on 278.

21 The opening exchange between hide and cobbler is in Sachs, *Rosshaut*, sig. Aii^r. The horse-hide's life story as a horse (versus what it experiences after it has been stripped from the dead horse's carcass) extends from sigs. Aii^r to Aiiii^r.

22 Sachs, *Rosshaut*, sig. Bii^r.

23 Horst Brunner, '"Ein liebhaber der poetrey": Umwelt und Dichtung des Hans Sachs', in Gabi Posniak (ed.), *Hans Sachs der Schuhmacher 1494–1576* (Offenbach am Main: Deutsches Ledermuseum/Deutsches Schuhmuseum, 1994), 32.

24 For the killing of horses, see Gisela Wilbertz, *Scharfrichter und Abdecker im Hochstift Osnabrück* (Osnabrück: Kommissionsverlag H. Th. Wenner, 1979), 59 and Jutta Nowosadtko, *Scharfrichter und Abdecker. Der Alltag zweier, unehrlicher Berufe' in der frühen Neuzeit* (Paderborn: Ferdinand Schöningh, 1994), 152. For selling the hides, see Doris Huggel, 'Abdecker und Nachrichter in Luzern, 15. bis 19. Jahrhundert', in Jürg Manser et al. (eds), *Richtstätte und Wasenplatz in Emmenbrücke (16.-19. Jahrhundert)*, 2 vols (Basel: Schweizerischer Burgenverein, 1992), ii. 204; Wilbertz, *Scharfrichter*, 52; Nowosadtko, *Scharfrichter*, 145.

25 For information on the activities and status of the Abdecker/Schinder/Wasenmeister in early modern Germany and Switzerland, see D. Huggel, 'Abdecker', 194–200 and 203–11; Wilbertz, *Scharfrichter*, 47–67 and 304–311; Nowosadtko, *Scharfrichter*, 118–161 and 354–64, and Hans Stampfli, 'Die Tierreste von Wasenplatz und Richtstätte', in Manser et al. (eds), *Richtstätte und Wasenplatz*, ii. 157–77.

26 Walter Bauernfeind, 'Löw' in Michael Diefenbacher and Rudolf Endres (eds), *Stadtlexikon Nürnberg* (Nuremberg: W. Tümmerls Verlag, 2000), 640. A number of secondary sources deal with animals in Nuremberg, but none of them mention

horses and deal instead primarily with dogs and pigs. See, for example, Corine Schleif, 'Who are the Animals in the Geese Book?' in Pia Cuneo (ed.), *Animals and Early Modern Identity* (Farnham: Ashgate, 2014), 209–42.

27   The book in question bore the title *Kunst puech von rossen, varben und kranckheiten.* See Wolfgang Milde, 'Das Bücherverzeichnis von Hans Sachs', in Merzbacher (ed.), *500 Jahre Hans Sachs*, 47.

28   See Gary Steiner, *Anthropocentrism and its Discontents: The Moral Status of Animals in the History of Western Philosophy* (Pittsburgh: University of Pittsburgh Press, 2005), 126–31. See also Erica Fudge, 'Two Ethics: Killing Animals in the Past and the Present', in The Animal Studies Group (ed.), *Killing Animals* (Urbana: University of Illinois Press, 2006), 101–2.

29   Thomas Aquinas, *Summa contra Gentiles*, III-II, q. 112 a. 12, cited in Steiner, *Anthropocentrism*, 128.

30   Steiner claims that Luther simply continued to uphold basic Thomist attitudes (131), but David Clough's work offers a more nuanced version of Luther's ideas. See David Clough, 'The Anxiety of the Human Animal: Martin Luther on Non-human Animals and Human Animality', in Celia Deane-Drummond and David Clough (eds), *Creaturely Theology: On God, Humans and Other Animals* (London: SCM Press, 2009), 41–60.

31   Christoph Jakob Lieb, *Practica et Arte di Cavalleria. Ubung und kunst des Reitens / in welcher der Bereuter / die Pferd nach ihrer Art und Natur zu unterweisen und abzurichten / erfahren und geuebt sein sol. Auch wie / und uff was weise dieselben in solcher handlung und abrichtung / zu schoenen wolstendigen Geberden und guten Tugenden sollen gewehnet und gezogen werden. Allen Liebhabern / und dieser Adelichen Ritterlichen Kunst zugethanen / zu sonderbahren Ehren und gefallen / uffs kuertze in zwey Theil verfast / und in offnen Druck gegeben* (Dresden, 1616), unpaginated (2).

32   See Pia F. Cuneo, '(Un)Stable Identities: Hippology and the Professionalization of Scholarship and Horsemanship in Early Modern Germany', in Karl A. E. Enenkel and Paul J. Smith (eds), *Early Modern Zoology: The Construction of Animals in Science, Literature and the Visual Arts* (Leiden: Brill, 2007), 339–59.

33   Sachs, *Rosshaut*, sig. Aii^v.

34   Lieb, *Practica et Arte di Cavalleria*, 52.

35   Sachs, *Rosshaut*, sig. Aiii^r.

36   Lieb, *Practica et Arte di Cavalleria*, 42–43. See also Lieb's book on bitting: *Gebissbuch* (Dresden, 1616).

37   Gabriel von Danup, *Idea oder Beschreibung eines Wolabgerichten Pferdes[...]* (Königsberg, 1624). Von Danup (1573–1629) notes his resignation in the unpaginated dedication to Maurice (Moritz), Prince of Orange (1567–1625); he states his age on p. 77. See also Koert van der Horst (ed.), *Great Books on Horsemanship: Bibliotheca Hipplogica Johan Dejager* (Leiden: Brill, 2014), 264–65. Von Danup had served as *Berreuther* for the Hohenzollern Georg Wilhelm (1595–1640), Calvinist Electoral Duke of Brandenburg and Duke of Prussia. See Andreas Gautschi and Helmut Suter, *Vom Jagen, Trinken und Regieren. Reminiszenzen aus dem Leben des Kurfürsten Johann Sigismund von Brandenburg, nach alten Briefen zitiert* (Limburg an der Lahn: C. A. Starke Verlag, 2005). Eventually, von Danup found employment under Christian IV, King of Denmark (1577–1648), who was the brother-in-law of Georg Wilhelm's father, Johann Sigismund (1572–1620). See Van der Horst (ed.), *Great Books*, 264.

38   [Von Danup], *Ein sonderliches Newes und Lesewuerdiges Gesprech welches gehalten ist worden fuer koenigl. Mayt: Apolline in Parnasso. Darinnen eingefuehret werden: Graff Wilhelm von Nassaw, Pater de Ney und Pirr Antonio di Ferrara wegen ubergebung einer klaeglichen Supplication der Pferde: Uber ihre gar zu Tyrannische Bereiter / und einem Gerichtlichem Abschied in der Sachen / daß / weil die bißhero begangene Fehler / nur*

allein aus mangel einer wahren definition eines wolabgerichten Pferdes / entstanden / Ihre koenigliche Mayt: allen Bereitern in Deutsch= Welsch=land und Franckreich / ernstlich aufferleget / sie sollen sich uber der wahren definition und beschreibung eines recht wol abgerichteten Pferdes / und was desselbigen nothwendige requisita weren / einigen und vergleichen (n.p., 1623).

39 Contrary to Van der Horst (ed.), *Great Books*, 264, von Danup's dialogue does not feature William Prince of Orange (1533–1584). The pamphlet clearly identifies the protagonist as 'Graff Wilhelm von Nassaw' and he is never referred to as the Prince of Orange. I believe we are dealing with Count Wilhelm Ludwig of Nassau-Dillingen (1560–1620), the Prince of Orange's nephew (son of Prince William's brother Johann VI) as well as his son-in-law (he married his uncle's daughter Anna!) The date of his death (1620) is closer to the date of the pamphlet's publication (1623), and the mediator in von Danup's dialogue is 'Pater de Ney' who is Johann Neyen, the Franciscan monk actively involved in Netherlandish politics in the opening decade of the seventeenth century, thus during Count William's lifetime and well after the death of Prince William. Von Danup's *Idea* (1624) is dedicated to Prince William's son, Maurice Moritz (1567–1625), captain general of the United Netherlands and Prince of Orange from 1618 to 1625. Maurice and Count William were both cousins and brothers-in-law. Both Count William and Prince Moritz were highly skilled and successful military men. 'Pirr Antonio di Ferrara' refers to the author of *Cavallo Frenato*, published in Naples in 1602. See Van der Horst (ed.), *Great Books*, 418–21; and Giovanni Battista Tomassini, *The Italian Tradition of Equestrian Art: A Survey of the Treatises on Horsemanship from the Renaissance and the Centuries Following* (Franktown: Xenophon Press, 2014), 161–74.

40 [Von Danup], *Ein sonderliches Newes und Lesewuerdiges Gesprech,* 2–5.

41 See Tomassini, *Italian Tradition,* 79–214.

42 [Von Danup], *Ein sonderliches Newes und Lesewuerdiges Gesprech,* 7.

43 Evidently this was the same in England. See Peter Edwards, *Horse and Man in Early Modern England* (London: Continuum, 2007), 32–3.

44 H. Häni, J. Lang, and G. Ueltschi, 'Ehemalige Richtstätte des Standes Luzern in Emmen (1562–1798) und dazugehöriger Wasenplatz (1562–1866): Pathologisch-anatomische Befunde am Tierknochenfundgut', *Schweizer Archiv für Tierheilkunde,* 136 (1994), 25. Of these, 130 could be identified according to sex: 20 stallions, 51 mares, and 59 geldings (H. Stampfli, 'Tierreste', 170).

45 Häni, Lang and Ueltschi, 'Ehemalige Richtstätte', 25.

46 Häni, Lang and Ueltschi, 'Ehemalige Richtstätte', 35.

47 Häni, Lang and Ueltschi, 'Ehemalige Richtstätte', 36.

48 Stampfli, 'Tierreste', 157.

Bibliography

Animal Studies Group (ed.), *Killing Animals* (Urbana: University of Illinois Press, 2006).

Anonymous, *Ein Newe und bewerte Rossartzney* (Strasbourg, 1583).

Armbruster, Karla, 'What Do We Want from Talking Animals? Reflections on Literary Representations of Animal Voices and Minds', in Margo DeMello (ed.), *Speaking for Animals* (New York: Routledge, 2013), 17–33.

Behr, Hans Joachim, 'Hans Sachs: Handwerker, Dichter, Stadtbürger. Versuch einer Würdigung anläßlich der 500. Wiederkehr seines Geburtstages', in Dieter Merzbacher (ed.), *500 Jahre Hans Sachs* (Wiesbaden: Harrossowitz, 1994), 9–16.

Brunner, Horst, '"Ein liebhaber der poetrey": Umwelt und Dichtung des Hans Sachs', in Gabi Posniak (ed.) *Hans Sachs der Schuhmacher 1494–1576* (Offenbach am Main: Deutsches Ledermuseum/Deutsches Schuhmuseum, 1994), 11–36.

Calarco, Matthew, *Zoographies: The Question of the Animal from Heidegger to Derrida* (New York: Columbia University Press, 2008).

Clough, David, 'The Anxiety of the Human Animal: Martin Luther on Non-human Animals and Human Animality', in Celia Deane-Drummond and David Clough (eds), *Creaturely Theology: On God, Humans and Other Animals* (London: SCM Press, 2009), 41–60.

Cuneo, Pia F. '(Un)Stable Identities: Hippology and the Professionalization of Scholarship and Horsemanship in Early Modern Germany', in Karl A. E. Enenkel and Paul J. Smith (eds), *Early Modern Zoology: The Construction of Animals in Science, Literature and the Visual Arts* (Leiden: Brill, 2007), 339–59.

———, 'The Artist, His Horse, a Print, and its Audience: Producing and Viewing the Ideal in Dürer's *Knight, Death, and the Devil* (1513)', in Larry Silver and Jeffrey Chipps Smith (eds), *Essential Dürer*, (Philadelphia: University of Pennsylvania Press, 2010), 115–29.

———, (ed), *Animals and Early Modern Identity* (Farnham: Ashgate, 2014).

———, 'Horses as Love-Objects: Shaping Social and Moral Identities in Hans Baldung Grien's *Bewitched Groom* (circa 1544) and in Sixteenth-Century Hippology', in Pia F. Cuneo (ed.), *Animals and Early Modern Identity* (Farnham: Ashgate, 2014), 151–68.

Danup, Gabriel von, *Ein sonderliches Newes und Lesewuerdiges Gesprech [...]* (n.p., 1623).

Danup, Gabriel von, *Idea oder Beschreibung eines Wolabgerichten Pferdes[...]* (Königsberg, 1624).

Deane-Drummond, Celia and Clough, David (eds), *Creaturely Theology: On God, Humans and Other Animals* (London: SCM Press, 2009).

DeMello, Margo (ed.), *Speaking for Animals: Animal Autobiographical Writing* (New York: Routledge, 2013).

Diefenbacher, Michael and Rudolf Endres (eds), *Stadtlexikon Nürnberg* (Nuremberg: W. Tümmerls Verlag, 2000).

Dinzelbacher, Peter (ed.), *Mensch und Tier in der Geschichte Europas* (Stuttgart: Alfred Kröner Verlag, 2000).

Donald, Diana, *Picturing Animals in Britain 1750–1850* (New Haven: Yale University Press, 2007).

Edwards, Peter, *Horse and Man in Early Modern England* (London: Continuum, 2007).

Enenkel, Karl A. E. and Paul J. Smith (eds), *Early Modern Zoology: The Construction of Animals in Science, Literature and the Visual Arts* (Leiden: Brill, 2007).

Freedberg, David, 'Empathy, Motion and Emotion', in Klaus Herding and Antje Krause Wahl (eds), *Wie sich Gefühle Ausdruck verschaffen: Emotionen in Nahsicht* (Berlin: Driesen, 2007), 17–51.

———, 'Memory in Art: History and the Neuroscience of Response', in Suzanne Nalbantian, Paul M. Matthews, and James L. McClelland (eds), *The Memory Process: Neuroscientific and Humanist Perspectives* (Cambridge: MIT Press, 2010), 337–52.

Freitag-Stadler, Renate (ed.), *Die Welt von Hans Sachs: 400 Holzschnitte des 16. Jahrhunderts* (Nuremberg: Hans Carl, 1976).

Fudge, Erica, *Animal* (London: Reaktion, 2002).

————, 'A Left-Handed Blow: Writing the History of Animals', in Nigel Rothfels (ed.), *Representing Animals* (Bloomington: Indiana University Press, 2002), 3–18.

————, 'Two Ethics: Killing Animals in the Past and the Present', in The Animal Studies Group (ed.), *Killing Animals* (Urbana: University of Illinois Press, 2006), 99–119.

Gautschi, Andreas and Helmut Suter, *Vom Jagen, Trinken und Regieren. Reminiszenzen aus dem Leben des Kurfürsten Johann Sigismund von Brandenburg, nach alten Briefen zitiert* (Limburg an der Lahn: C. A. Starke Verlag, 2005).

Häni, H., J. Lang, and G. Ueltschi, 'Ehemalige Richtstätte des Standes Luzern in Emmen (1562–1798) und dazugehöriger Wasenplatz (1562–1866): Pathologisch-anatomische Befunde am Tierknochenfundgut', *Schweizer Archiv für Tierheilkunde,* 136/1 (1994), 24–37.

Hediger, Ryan, 'Our Animals, Ourselves. Representing Animal Minds in *Timothy* and *The White Bone*', in Margo DeMello (ed.), *Speaking for Animals* (New York, Routledge, 2012), 35–47.

Herding, Klaus and Antje Krause Krause Wahl (eds), *Wie sich Gefühle Ausdruck verschaffen: Emotionen in Nahsicht* (Berlin: Driesen, 2007).

Huggel, Doris, 'Abdecker und Nachrichter in Luzern, 15. bis 19. Jahrhundert', in Jürg Manser et al. (eds), *Richtstätte und Wasenplatz* (Basel: Schweizerischer Burgenverein, 1992), 193–221.

Karant-Nunn, Susan, *The Reformation of Feeling: Shaping the Religious Emotions in Early Modern Germany* (Oxford: Oxford University Press, 2010).

Lieb, Christoph Jakob, *Practica et Arte di Cavalleria. Ubung und kunst des Reitens [...].* (Dresden, 1616).

Manser, Jürg et al. (eds), *Richtstätte und Wasenplatz in Emmenbrücke (16.-19. Jahrhundert)*, 2 vols (Basel: Schweizerischer Burgenverein, 1992).

Martini, Wolfram, Jochen Küppers and Manfred Landfester, 'Römische Antike', in Peter Dinzelbacher (ed.), *Mensch und Tier in der Geschichte Europas* (Stuttgart: Alfred Kröner Verlag, 2000), 87–144.

Merzbacher, Dieter (ed.), *500 Jahre Hans Sachs: Handwerker, Dichter, Stadtbürger* (Wiesbaden: Harrossowitz, 1994).

Milde, Wolfgang, 'Das Bücherverzeichnis von Hans Sachs', in Merzbacher (ed.), *500 Jahre Hans Sachs*, 38–55.

Nalbantian, Suzanne, Paul M. Matthews and James L. McClelland (eds), *The Memory Process: Neuroscientific and Humanist Perspectives* (Cambridge: MIT Press, 2010), 337–52.

Nowosadtko, Jutta, *Scharfrichter und Abdecker. Der Alltag zweier, unehrlicher Berufe' in der frühen Neuzeit* (Paderborn: Ferdinand Schöningh, 1994).

Posniak, Gabi (ed.) *Hans Sachs der Schuhmacher 1494–1576* (Offenbach am Main: Deutsches Ledermuseum/Deutsches Schumuseum, 1994).

Rothfels, Nigel (ed.), *Representing Animals* (Bloomington: Indiana University Press, 2002).

Sachs, Hans, *Die ellend klagend Rosshaut* (Nuremberg, ca. 1557).

————, *Werke*, eds Adelbert von Keller and E. Goethe, 26 vols (Tübingen: H. Laupp, 1870–1908).

Schleif, Corine, 'Who are the Animals in the Geese Book?' in Pia Cuneo (ed.), *Animals and Early Modern Identity* (Farnham: Ashgate, 2014), 209–42.

Silver, Larry and Jeffrey Chipps Smith (eds), *The Essential Dürer* (Philadelphia: University of Pennsylvania Press, 2010).

Stampfli, Hans, 'Die Tierreste von Wasenplatz und Richtstätte', in Manser et al. (eds), *Richtstätte und Wasenplatz*, 157–77.

Steiner, Gary, *Anthropocentrism and its Discontents: The Moral Status of Animals in the History of Western Philosophy* (Pittsburgh: University of Pittsburgh Press, 2005).

Strauss, Gerald, *Nuremberg in the Sixteenth Century* (Bloomington: Indiana University Press, 1976).

Theiß, Winfried, *Exemplarische Allegorik. Untersuchungen zu einem literarischen Phänomen bei Hans Sachs* (Munich: Wilhelm Fink Verlag, 1968).

Tomassini, Giovanni Battista, *The Italian Tradition of Equestrian Art: A Survey of the Treatises on Horsemanship from the Renaissance and the Centuries Following* (Franktown: Xenophon Press, 2014).

van der Horst, Koert (ed.), *Great Books on Horsemanship: Bibliotheca Hipplogica Johan Dejager* (Leiden: Brill, 2014).

Waal, Frans de, *The Age of Empathy* (New York: Three Rivers Press, 2009).

Wilbertz, Gisela, *Scharfrichter und Abdecker im Hochstift Osnabrück: Untersuchungen zur Sozialgeschichte zweier, unehrlicher' Berufe im nordwestdeutschen Raum* (Osnabrück: Kommissionsverlag H. Th. Wenner, 1979).

Part II

Use and abuse

4 The tale of a horse

The Levinz Colt, 1721–29

Peter Edwards

The varied essays published in this volume indicate that the field of Animal Studies is a very broad one, a quality that encourages an inter disciplinary approach to subjects included within its compass. It has benefited my own research, as reflected in the comparison of the books I wrote on the subject in 1988 and 2007.[1] My training as an empirical social and economic historian is clearly evident in the strictly anthropocentric standpoint I took in the first book, *The Horse Trade of Tudor and Stuart England,* which deals with horses as a commodity to be bought, sold, and used solely in the interests of humans. Whilst my basic stance remained the same, the approach softened by the time I published the second book, *Horse and Man in Early Modern England,* in 2007. A more nuanced account of the relationship, it recognized that horses were intelligent, sentient creatures, who interacted with their environment and could even change it through the exercise of agency. In between, the contents of a number of articles and essays indicate that this was no sudden Damascene conversion but the result of thirty years of exposure to the work of scholars from other disciplines with different approaches to and views on the nature of the relationship between humans and the rest of the natural world.

As the title of this book stresses, the relationship between the two species is a two-way process. However, to examine the historical relationship from the animal's perspective is difficult, for as Erica Fudge succinctly puts it, animals 'had no voices and left no textual traces'.[2] One possible way forward is the construction of animal biographies (wherever sufficient evidence exists), a point Fudge raised in 2004 when questioning the absence of animal profiles in the Dictionary of National Biography.[3] She cites the examples of Morocco the Intelligent Horse, a celebrity for over twenty years at the turn of the sixteenth century, and Blind Bess, a famous bear, who was baited at the Hope Theatre in London for a number of years in the mid-seventeenth century. As micro-historians argue, reducing the focus of the research to a detailed small-scale study can offer general insights too. In essence, this is the aim of this essay, which focuses on the Levinz Colt, bred and reared at Lord Edward Harley's stud at Welbeck (Nottinghamshire) in the 1720s. The stud, once the possession of the famed horseman, William Cavendish, 1st duke of Newcastle, devolved to Harley's wife, Henrietta, on the death of her father,

John Holles, the duke of Newcastle (second creation), in July 1711. Nonetheless, she only gained de facto possession of it in December 1716 on the death of her mother, Countess Margaret (*née* Cavendish), who had bitterly opposed the terms of her husband's will.[4] Fortunately, we can follow the progress of the Levinz Colt through a set of over 400 letters written to Harley by his local agent, Isaac Hobart, between August 1721 and September 1733. Of course, the letters give a lop-sided account, one taken from the human perspective and not direct from 'the horse's mouth', as it were. As the collection only contains three letters from Harley to Hobart neither do we hear his master's voice, although we can access it indirectly through Hobart's responses to his instructions. Likewise, the correspondence allows us a glimpse, if a shadowy one, of the relationship from the perspective of the Levinz Colt and other horses at the stud.

I

On 4 April 1723 Hobart informed his master that, 'We have this morning a large colt foal from the whitefoot or Levinz mare'. He described him as dark chestnut in colour, with four white legs and a half star.[5] Hobart thought so highly of the colt as he developed that on 17 March 1729 he recommended the grey horse to Sir Edward, who became 2nd earl of Oxford in 1724, as his personal mount. From these two letters we already know of the horse's parentage, his strength (Harley was growing portly), and by inference his temperament, the networks that facilitated the process and the fact that his coat-colour changed over time.[6] Other references fill out the picture: we discover when he was broken-in, the treatment he received for his ailments, his designated role and monetary worth, and how Harley and his staff (and other breeders) viewed him. Naturally, Hobart discusses other horses at the stud during the period covered by the letters, especially in the first few years when Welbeck Abbey housed the main stud on the estate. Even the tailing off of information, reflecting the rationalization of the enterprise at Harley's chief residence at Wimpole (Cambridgeshire), provides evidence of this shift in policy. Thus, the letters written by John Cossens, Hobart's counterpart at Wimpole, help to sharpen the picture, as do those written by numerous other correspondents to Harley.[7] An incomplete account book, dated 1724–41, provides additional information on income and expenditure on the stud.[8]

   The documentation enables us to look at the Levinz Colt as an individual, something most historians of pre-modern animals cannot do, partly because no other species enjoyed the same relationship with humans or occupied such an exalted position in the animal kingdom. In 1570 Conrad von Heresbach had asserted that 'the Horse may worthiest challenge the chiefest place, as the noblest, the goodliest, the necessariest, and the trustiest beast that wee vse in our seruice'. Over two centuries later, John Lawrence was making the same point, declaring that horses were the most beautiful of all four-footed animals and superior to all others in terms of the symmetry of their bodies, their speed and in their general utility to mankind. Moreover, in line with a growing awareness of animal physiology and intelligence over the course of the intervening period, he added that the horse 'possesses in common with the human race the reasoning faculty, the difference consisting only in degree, or quantity'.[9]

If many pets enjoyed a special relationship with their owners, none of them possessed the same combination of physical and non-physical attributes or the same iconic appeal. A letter written in December 1705 by Harley's uncle Nathaniel, then a factor in Aleppo, makes the point. In it, he confessed to Edward, his own brother, that he was finding it difficult to obtain a suitable horse for him, 'There being something so peculiar in everyone's fancy That next to choosing a wife for a man The most difficult thing is to please him with a Horse'.[10] After all, a rider and his favourite horse spent a good deal of time together – perhaps more than a husband did with his wife – and, once established, the bond between them was so strong that it might appear like a marriage. Indeed, the owner might attach anthropomorphic qualities to his mount. According to Margaret, the wife of William Cavendish, her husband had so great a love of his horses that this affection was reciprocated: 'they seemed to rejoice whensoever he came into the Stables, by their trampling action, and the noise they made... and when he rid them himself, they seemed to take much pleasure and pride in it'.[11] From this observation, one might conclude that the horses were exhibiting such abstract human characteristics as pleasure, pride, loyalty, faithfulness, and a desire to please. If it contains an element of wish-fulfilment, the narrative revealing more about the writer's view of her husband's prowess as a horseman than the horses' response to him, it does suggest that their relationship with him was a mutually satisfying one.[12]

Significantly, the example refers to horses who were being trained for the *manège*, a set of exercises akin to the modern dressage, a pursuit in which Cavendish excelled. As it required years of training and close contact between horse and rider, the horses would certainly recognise Cavendish and respond to him. Moreover, as the aim was to achieve complete harmony between them, characterised by the concept of the centaur, there had to be a synchronisation of thought processes as well as of physical actions. Only then could horse and rider act in concert as a single unit. But it was a two-way process, requiring an effort of comprehension by the rider and horse alike. While the horse had to learn and respond to the rider's instructions, the rider had to work with the horse, along the grain of the creature's nature if he were to succeed. Clearly, such horses – and, by extension, other horses – were not the mindless automatons that Descartes would have us believe. As William Cavendish remarked, 'if he [the horse] does not think (as the famous philosopher DES CARTES affirms of all beasts) it would be impossible to teach him what he should do'.[13] After one of Cavendish's displays of horsemanship in Antwerp in the 1650s noble observers told the governor, Don John, that the horses he rode 'wanted nothing of Reasonable Creatures, but Speaking'.[14]

Horses kept by owners, who shared Cavendish's approach and who had spent a good deal of time with them from birth, would have established a similar positive relationship. Harley was not such a person. He lived at Wimpole and left the care and maintenance of his horses to Hobart and to his stud-master, John Bowron, and his staff. Harley's relationship with the horses began after they had been broken-in and trained. In general, therefore, the first humans whom the horses met and with whom they established relations were grooms and stable lads. In close contact with horses on a daily basis, they might form a bond with their charges and look after them well. On the other hand, as it was onerous job, often sleeping in the stable loft above the horses, being constantly on call and spending their time carrying out dirty, menial tasks, they might act in a casual way, even mistreating the horses. At least the horses at Welbeck had two hard-working, dedicated officials in Hobart and Bowron to protect their interests. Indeed, on 15 August 1722, Hobart informed Harley that the stable lads, Ralph and Robin, were unreliable, citing their absence several nights a week.[15] Unfortunately, Hobart had trouble finding suitable replacements. As he wrote on 25 September 1723, 'We very much want a sober careful groom to be always on the spot here, who will stick close to business. Such a one is not to be had in this country. The person I hired knows nothing of the matter and if he was ever good for ought, has now learn'd to be otherwise, but I cannot do at present without him'.[16]

II

The Levinz Colt certainly possessed exceptionally good genes. His dam, the Levinz or Whitefoot mare, was born nearby at Harley's father-in-law's stud at Haughton, probably in 1707. In a list of brood mares there in 1712, she is described as Sorrel Whitefoot, five years old, got by the Paget Arab out of a dam got by Hoboy out of a dam 'extraordinary fine out [of] Fenwicks Breed'. Another document adds that Sorrel Whitefoot's dam was the Darcy Whitefoot Mare. The duke had obtained her sire, Paget's Arabian, from William, 6th Lord Paget, ambassador to the Ottoman Empire between 1692 and 1701, on his return to England in 1703. Her dam was bred at James Darcy's famed stud at Sedbergh (North Riding).[17] Indicative of the standing and distinction of this establishment was the agreement made between Charles II and James Darcy the elder, newly appointed as Master of the Royal Stud, in 1660 to deliver 'twelve extraordinary good Colts' annually to the king for a fee of £800. The dams were termed 'Royal Mares', a name that was also attached to surviving mares of the disbanded royal stud at Tutbury (Staffordshire). Because of the quality of Darcy's royal mares, they and their offspring were much in demand by aristocratic breeders. Indeed, the mares made an important contribution to the genetic make-up of the English thoroughbred.[18]

The Levinz Colt's sire, the Bloody Shouldered Arabian, was an even finer animal.[19] Nathaniel Harley had acquired him as a two-year-old in 1715 but had only managed to slip him out of Aleppo in January 1720. Even with his experience in obtaining and shipping out Arabians, this was no mean feat. Apart from Arab reluctance to part with their horses, the Ottoman government banned the export of such a valuable commodity and watched out for smugglers. On 15 May 1717, Nathaniel wrote that the horse belonged to 'a celebrated Race among the Gordeens, from whom 'tis difficult to procure a Horse, the Race in a manner lost among them. Has no great speed but is strong and would serve well for the *manège* being easy to learn anything'. On 6 January 1720, Nathaniel finally informed his brother that he had shipped out the horse, whom he described as, having 'a good body, a long well shaped Neck, a pair of Glistening Eyes, and stands upon four good leggs'.[20]

This discussion of the Levinz Colt's lineage raises some important issues. First, it reveals the growing interest that elite breeders were taking in the genealogy of their horses, a practice that the Arabs had been following for hundreds of years. Commenting on the valuable racehorses he saw at Penkridge (Staffordshire) in the 1720s, Daniel Defoe observed that they were 'famous for the breed, and known by their race, almost as well as the Arabians know the genealogy of their horses'.[21] On 22 July 1722, Harley specifically asked Hobart to provide genealogical details of his horses: 'I would know the Number of Mares I have the Colts and by what Horses got the Age of the Colts their marks and how they are Disposed that is the grounds where they are and how likely to prove'.[22] Noting a horse's pedigree originated as a guide to breeding racehorses, but the custom spread to other types of horses as well. Economics played a part in this development. Because of the cost of maintaining a racing stud, owners earned money by hiring out their stallions to all who could afford their services, for whatever purpose. As a result of cross-breeding, the gentry improved their stock of horses, especially saddle mounts (including those for hunting and the cavalry).[23] Newspapers of the time are full of advertisements extolling the pedigree and potency of the stallions.[24]

The Levinz Colt's pedigree, as recorded by Hobart at the horse's birth, also demonstrates the existence of networks among elite breeders.[25] They had long formed such associations, especially in areas noted for the production of high-quality horses, but they benefited from the greater availability of genealogical information from the post-Restoration period onwards. It enabled them more effectively to pick and choose a horse's blood-lines. Thus, Harley regularly received requests from his peers to allow his stallions – the Dun Arabian and, then, the Bloody Shouldered Arabian – to cover their mares. Among the brood mares awaiting the services of the Bloody Shouldered Arabian at Welbeck on 22 May 1723 were two belonging to the marquis of Carmarthen, three of the earl of Kinnoul's, two of Mr Levinz's and one each owned by Mr Chetwynd, Mr Pelham and Mr Snell.[26] In turn, Harley's mares received outside stallions. Among the colts stabled at Welbeck on 9 August 1722 were those sired by stallions belonging to the earl of Kinnoul and the duke of Bolton out of the Whitefoot and Young Arabian Mares respectively.[27]

Because the elite valued horses for their symbolic as well as their functional qualities, they were obsessed with fashion and outward appearance: breed, conformation, size, action, and even colour. These concerns are reflected in their horses' nomenclature. The names of Harley's horses, for instance, invariably fall into the three most common categories: the person who bred or sold the horse; the horse's pedigree; and coat colour with markings. An indication of size was often added.[28] The first two categories provided a measure of quality control, important to elite horsemen, who by the 1720s were becoming almost as concerned with the genealogy of their horses as they were about their own family. To have a Darcy royal mare and a Fenwick mare in the Levinz Colt's bloodlines was clearly a sign of good breeding. As the Levinz name was taken from his dam, this suggests that William Levinz of Grove Hall (Nottinghamshire), a political associate of Harley's as well as a member of his horse-breeding circle, had owned the mare at one time. Indeed, some of his horses may well have had the appellation 'Harley' as the Bloody Shouldered Arabian regularly covered his mares.[29] Several horses are called 'Arabian', an indication of the value that contemporaries placed on the breed. The names of two of Harley's horses – the Chestnut Snake Colt and the Careless Colt – indicate descent from notable racehorses. Because racehorses had to be individually identified, they needed a unique name, although the system did not become fully operational until well into the eighteenth century. This requirement allowed owners the opportunity to exercise their creative ability with many of the horses given names indicating potency, a wry sense of humour or the benefits of a classical education.[30]

The third category – coat-colour plus markings – was common partly because it was distinctive: the large reddish stain on a shoulder of the Levinz Colt's sire, the Bloody Shouldered Arabian, brooked no alternative name. Markings were also useful in identifying a horse's genealogy. The pedigree of the colt's dam, the Whitefoot Mare, for instance can be worked out from the list of horses at the Haughton stud of Harley's father-in-law. That said, there is an alternative genealogy.[31] The highlighting of coat colours also reveals the continuing influence of astrology and the humoral theory on contemporary thought, which linked coat colour with a horse's temperament. In *Cauelarice* (1607), Gervase Markham emphasized the relationship, listing horses of particular coat colours in one or another of the four humoral categories of melancholic, phlegmatic, sanguine, and choleric. If Markham's assessment is correct, the Levinz Colt's temperament changed dramatically from being lazy, dull, and lacking in spirit (phlegmatic) to bold, nimble, good natured, and free spirited (sanguine).[32] By the 1720s, however, opinion had shifted. In the year that the Levinz Colt was conceived (1722), William Gibson, a seminal writer on farriery, was dismissive of the theory: 'It will be very little to our purpose to spend much time about its [hair] Production, or how it comes to be of so many different Colours'. In the 1720s, however, fashion could still dictate choice, whether for a matching team of coach horses or a strikingly coloured saddle horse.[33]

III

Unfortunately for Harley, the Levinz Colt was the last foal that the Whitefoot Mare produced. She received the stallion shortly after the colt's birth, but on 24 December 1723, Hobart reported that she had a surfeit and was very sick. He added that if she had been in foal, she had probably aborted the foetus. Although the mare was still ill in February and March 1724, another attempt was made that year. She did conceive but had lost her foal by 29 March 1725, even if Hobart could not explain why: 'she was in good order and well to all appearance but the day before'.[34] Within a few days, Harley had told Bowron not to have her covered again. On hearing the news, Hobart hesitated, arguing, 'I am sure it can be no prejudice to cover her'. He did have a buyer, a Lincolnshire gentleman who had come to Welbeck to view the mare the previous week, but he must have fobbed him off. As Hobart admitted, 'I presumed your Lordship wou'd not part from her being the only strong mare of thorow blood in the Studd', a comment that emphasizes the Levinz Colt's impressive lineage. Surprisingly, Harley eventually agreed and on 17 May 1727, Arthur Duchford's stallion served her and two other brood mares. She did not conceive.[35] At the time, Harley was deeply in debt and was trying to retrench. Unlike his passion for collecting books and manuscripts, maintaining a stud was merely an interest and therefore easier to jettison.[36] It is also significant that the timing of his decision to run down the stud coincided with the failure of his prize mare to conceive. Conversely, Hobart must have hoped to squeeze another foal out of the mare, aware of Harley's intentions and the certainty that he would not provide him with enough money to buy a suitable replacement.[37] Although she was not an active agent in the process, the question of the Whitefoot Mare's fertility had a profound effect on the fortunes of the stud. Just as the presence of such a well-bred brood mare at Welbeck contributed greatly to Harley's decision to revive the stud, so her failure to conceive was instrumental in its demise.

Surprisingly, given their concern for the quality of their horses, the elite expected their brood mares to produce a foal every year. Commentators disapproved of the custom, which, because the gestation period for a foal was a full eleven months, gave the mare little time to recover.[38] Harley, like many gentlemen-breeders, did not pay attention, prioritizing economics over best practice. Of all his brood mares, the Whitefoot Mare had the best pedigree and was therefore the one expected to produce the most sought-after foals. But, as she suffered from an unknown genetic weakness,[39] annual pregnancies affected her health and damaged some of her offspring. On 15 August 1722, Hobart was worried that the two-year-old Grey Colt and the three-year-old Dun Filly had inherited the mare's infirmities – and with reason.[40] After failing to conceive in 1720 she produced fine foals in 1721, 1722 and 1723 before miscarrying later in 1723 and in 1724. The Betty Mare also aborted her foal in 1724.[41] If brood mares were being treated as if on a production line of foals, they could and did exercise a degree of agency by refusing the stallion, as the Old Arabian Mare did in July 1724. Hobart feared that they would lose her in the winter. If this mare was ageing, stud masters often had to encourage other, younger mares to accept the designated stallion by employing a teaser.[42] On 1 May 1723, Hobart nominated the dun colt brought from the North for the task, as Bowron needed a horse to 'try' the mares. Condemning the horse to a life of frustration, Hobart pointed out, 'Truly there is nothing in him promising'.[43]

IV

Horses were notoriously susceptible to illness and injury, but at least those kept by the elite led a far more cosseted life than those maintained by the population at large. Their owners housed them in imposing stable blocks, hired staff to tend to them, and restricted them to a specific task.[44] When they were sick or injured, they were rested and given the best treatment available at the time. At Welbeck, Hobart's reports reveal that Harley's horses regularly needed attention, sometimes several of them at once. On 14 September 1723, five of the horses were ill: two lame, one almost blind, one with a cough, and one that was almost blind and had a cough.[45] The horses also suffered from the epidemics that periodically swept the equine population of England. In October 1727, for instance, all the horses at Welbeck, including the Levinz Colt, were seized by 'the most violent coughs', a sickness that, according to Hobart, had affected every horse in the country.[46] Bowron could deal with minor ailments; for help he could read the farriery section in horsemanship manuals, which appeared in growing numbers from the late sixteenth century onwards.[47] At Welbeck, treatment was still based on the Galenic humoral theory, which meant that an imbalance in one of the four humours had to be rectified by bleeding, purging, and the application of poultices. For more serious complaints, he called in the estate's contract farriers, Mr. Marriott, father and son or, if at Wimpole, Thomas Darlow, then John Elsom.[48] On 14 March 1725, the Levinz Colt, then at Wimpole, was very lame. However, Bowron could not find Elsom to give the colt treatment. Perhaps as a result, the injury reoccurred the following year. On 21 May, he was reportedly in such a bad way that Bowron decided to take the colt to old Mr Marriott's home 25 miles away at Brodsworth as soon as the animal was fit to travel. The colt left in mid-July and did not return until 10 September.[49] On 29 September, Hobart paid Marriott's bill of 8s. 2d. for 'firing' (cauterizing) the colt, a common treatment for lameness at the time.[50]

V

Breaking-in clearly marked an important step in the relationship between a horse and his owner, as it emphasized the control of one species over another, perhaps after a struggle to determine mastery. At least, the methods used softened over the course of the early modern period as owners – those who had read the manuals on horsemanship – absorbed the view of the writers that the process should be carried out sympathetically rather than by beating the animal into submission.[51] One hopes that the specialist horse breakers used by the elite had read the manuals too.[52] At Welbeck, young horses like the Levinz Colt tended to be broken-in during their fourth year, although some commentators preferred breeders to wait a couple of years longer, that is, when they had reached maturity: fillies at the age of five and colts a year later.[53] Writing in 1674, John Halfpenny contrasted late-trained colts, which possessed well-knit joints and sinews, tough hooves, and good eyesight, with those broken in at the age of two or three with their poor eyesight, brittle hooves, and weak backs. On the other hand, Michael Baret, Gervase Markham, and William Cavendish thought that fillies and colts should be broken in when two and three years old, respectively.[54] Bringing the colts and fillies indoors in wintertime had already accustomed the horses to humans and that process of acclimatization continued up to the point of mounting the animal. Gervase Markham allowed two weeks for the breaking-in process, gradually introducing the saddle and by degrees mounting the horse.[55] Even after the horse had accepted the rider, patience was required. On 30 May 1726, for example, Thomas Harley, his father's cousin, wrote to Harley from Brampton Bryan (Herefordshire), the family's ancestral home, to offer him a young mare 'fitter for your stable then to be spoiled here'. Rogers had ridden her but as she was very gentle, she would have to be out to graze 'after this first rideing to prevent a fret', should Harley not send for her immediately.[56] Similarly, in June 1728 Hobart sought to ease in the filly belonging to Harley's daughter, Margaret. Because the filly had been ridden gently, she had 'a pretty good mouth', but as she was young and small, Hobart decided that she would not be ridden again that season.[57]

However carefully the breakers carried out their work, unbroken horses were not necessarily willing to submit to the bit and bridle, or to having people on their backs. They resisted, to the point of unseating their rider, an occurrence so common that the manuals devoted a considerable amount of space to dealing with a difficult horse. When on 28 September 1726 Hobart informed Harley that three colts, including the Levinz Colt, were coming four years old and 'would have them broke forthwith' he must have worried how three 'headstrong' and 'very strong' young horses would respond.[58] As John Ratcliffe had successfully broken-in the three four-year-old fillies in the spring, Hobart hired him to ride the three colts the following February. He earned £9 3s. for breaking-in the seven horses.[59] Hobart was even more concerned about the Careless Colt, born in 1726 but already proving a handful by the time he was three years old. On 15 March 1729, Hobart asked Harley if he could sell the colt as 'he grows so Rampant'. At the end of May, work connected with the 'water engine' so distracted the colt that 'he fret's himself almost to death', leading Hobart to declare that 'he will be so rude that it will be difficult to break him'. Indeed, on 7 July Hobart advised Harley to sell him unbroken.[60] Here, Hobart is investing the colt with the attitudes of a badly behaved adolescent.[61]

VI

Hobart paid particular attention to the Levinz Colt because he felt the horse possessed the qualities of strength, speed, and good blood that his master sought. Described as large at his birth, he was included in the short list of horses deemed suitable for Harley's service on 2 May 1724. Three weeks later Hobart expressed his pleasure that Harley liked the young grey colt, and promised to take good care of him.[62] As a strong 'thorough-pad', a high-quality, easy-paced saddle mount, he was the type of horse that Harley liked to ride as he was growing portly with age. These horses had become fashionable in the post-Restoration period but good specimens were hard to find because they had to be 'strong, light and nimble', a rare combination of qualities. In 1680, the earl of Halifax complained that 'it is almost as possible to get a horse that flyeth as a horse that paceth, I mean one that doth it well, so rare that kind of creature is grown amongst us'.[63] Good specimens were therefore expensive. In a letter written to Harley on 7 August 1730, Morgan Vane, a kinsman of Harley's wife, valued a plain strong-pacing stallion at 100 guineas and a very plain strong-pacing gelding at 40 guineas but did not price a lesser pad.[64]

This was the market Harley hoped to exploit, and this meant that skilled pacers, specialists who trained horses in this gait, were in great demand.[65] But because there was money to be made, it paid members of the elite to hire one on recommendation. Thus, on 2 July 1723, Hobart reported that the pacer recommended by William Levinz had started to pad three of the stud's horses. He left almost six weeks later, claiming to have made three very good pads. Hobart was less certain, only admitting that they moved like pads and travelled at about four miles per hour without tripping or starting.[66] As image mattered, only horses who had perfected the gait and looked good fetched premium prices. In September 1726, Hobart also had difficulty finding a proficient pacer to train the Levinz Colt and the two other three-year-olds once they had been broken-in, complaining that he would 'want a skillful & carefull Person to ride them afterwards which is not to [be] had here'.[67] So when Mr Waite of Richmond (North Riding) had tried to sell Harley two pads earlier in the year, he may well have told the truth when he declared that he had been 'at great charge in pacing and setting them to the road'.[68] In November 1731, Morgan Vane seems to have found the perfect, well-trained pad for Harley in London, declaring that he could not acquire a horse 'more fit for your purpose'. A fortnight later, having just ridden the horse, he declared him to be 'one of the pleasantest surest horses I ever yet saw very nimble, quick & never starts, he is very light in the hand & firm in the mouth'.[69] As Vane's description indicates, horses were sentient beings, possessed of their own individuality and temperament. To establish a good relationship it was important therefore that horse and rider were temperamentally suited to each other.

The importance of temperament in determining the fortunes of Harley's horses can be illustrated by comparing the experience of the Levinz Colt with that of the Careless Colt. Both had exceptional pedigrees, were fine horses, and physically met Harley's need for strong, well-proportioned pads. Both showed initial promise; as yearlings they were both deemed fit for Harley's service, while Hobart also assessed the Careless Colt as very fine.[70] Both had suspect temperaments, however. On 27 January 1729, Hobart told Harley that he feared that the Levinz Colt would be 'vicious', perhaps an even more damning indictment than his description of the Careless Colt as 'rampant' in March 1727.[71] Both might have been sold as young horses: the Levinz Colt to Mr Chaworth of Annesley Hall (Derbyshire) as a stud stallion in October 1726 and the Careless Colt to an unknown buyer who had offered sixty guineas for him in March 1729.[72] At that point, the Careless Colt looked the better bet, especially as the Levinz Colt's action was poor: on 27 January 1729, Hobart stated that 'I cannot think that the Grey horse will please your Lordship upon the road or any where Else: He trots high, and when he Gallops will have his head down'. Even so, he rejected the marquis of Carmarthen's bid of 100 guineas for him. On 10 March, Hobart reported that the marquis, hearing that Harley was disposing of horses, had asked to have first refusal of the Levinz Colt.[73] At the same time Hobart was trying to sell the Careless Colt, but with no success. He therefore had him broken-in. William Wilkinson, hired for the purpose, evidently did a good job, for on 24 November Hobart reported that 'the colt is just broke and I think very well, he has no vice at present; I hope he will make a usefull horse'.[74] In the meantime, Hobart's confidence in the Levinz Colt had been vindicated for on 17 March 1729 he could state that, 'I hope the grey Horse will carry your Lordship with safety. I am sure he is strong enough, and the Groom tells me he is well temper'd'.[75] The Careless Colt, on the other hand, had regressed by the following April as a result of a chafing martingale, which caused 'a flux of humours to fall into his chest and body' and quickly reduced him 'to a most miserable low condition'. Although he had recovered, the illness had reactivated his vicious streak.[76] In June 1733, Hobart had him gelded, partly to render him more docile and partly as a precaution against the passing on of his 'bad blood' to his descendants. In August 1734, Mr Joseph Bright paid six guineas for the horse, 'the vitious Lame Careless Gelding, not worth one farthing'.[77]

VII

Although Harley's motives for restoring the stud at Welbeck are uncertain, he expected it to produce strong, finely proportioned pads for his own use. Surplus horses or those that did not match his exacting requirements might be ridden by his officials or sold off, either privately to his peers or publicly at fairs. Harley's decision had spatial consequences too. As Pearson and Weismantel have pointed out, animals had an impact on their physical environment.[78] At Welbeck, once Harley gained control of the estate in the winter of 1716–17, he sacked the shepherd and sold off his flock of 1000 sheep, 700 of which were being kept in demesne closes in March 1717.[79] The other 300 were roaming in the forest, which if left uncropped would have reverted to scrub. The demesne closes had to be improved too because fine horses had to be kept in well-maintained paddocks and over-wintered on good quality hay, augmented with corn rations. On 12 August 1721, Hobart told Harley that he required an additional paddock, which entailed draining some boggy ground to compensate for the loss of hay in the Stand Close, the pasture earmarked for upgrading. In March 1724, he was asking for another paddock, should Harley want to keep a stud at Welbeck. Even so, he had to rent a paddock from Mr. Porter the following year, where he kept six colts, including the Levinz Colt, for six weeks.[80]

With horse paddocks at his disposal, Hobart could not only ensure the quality of the grass but also supervise the placement and movement of the horses, so essential for a well-run elite stud. He could control the breeding process, separate the brood mares and their foals from the rest of the herd, and isolate problem horses like the Careless Colt. On 5 April 1729, Hobart expressed concern for that colt because the only place away from other horses was unsafe.[81] Keeping pregnant mares, foals, and young horses separate from other horses prevented accidents as well as attacks from other horses. As Sir Roger Pratt of Ryston Hall (Norfolk) noted in his commonplace book on 10 July 1680, heavily pregnant brood mares should be kept away from others in the herd to avoid the danger of miscarrying if kicked. Two years later a young filly foal running with her dam in a 'rabble' of horses died after breaking her leg.[82] At Welbeck, Hobart had to inform Harley on 25 February 1727 that the Arabian Colt had received a kick or a bite from one of the other colts, possibly from the Levinz Colt or Bay Colt, the other two three-year-olds. Although serious enough to put his life at risk, the colt had made a full recovery.[83] It was essential that the fencing be strong enough to stop the horses breaking out, causing damage, flattening growing crops, or wantonly impregnating brood mares. So, trees had to be cut down for paling. In November 1730, Welbeck servants earned a total of £2 12s. 5d. for a pale fence between Stand Close and Harlot Lane: 2s. 6d. to Joshua Newet for stubbing and butting two trees for pales; 5s. 5d. to Robert Shepherd for making 480 pales and £2 4s. 6d. to John Presley for setting up sixteen acres of pale fence and other carpentry work.[84]

Among the most important issues that Harley had to consider was the cost of fodder, as the management of the Levinz Colt illustrates. In the first few months of his life, he ran with his dam and depended on her milk. On 17 August 1723, he could be found in Stand Close, the new paddock, along with his dam, five other brood mares and two foals. Five older colts were grazing in the Toad Home and Paddock.[85] Although commentators recommended prolonged suckling because it enhanced the strength and wellbeing of a foal and extended his or her life, it was customary to wean foals in the autumn. That winter foals fed on hay and oats, perhaps with a little beans or peas. Hobart is silent on the issue of a foal's ration, but the records of Harley's father-in-law reveal that at Haughton in 1708 foals and yearlings received two pecks of oats a week in the stable.[86] As they grew, the foals' rations increased. A letter of 5 June 1729 from Cossens indicates the upper limit, when he responded to Harley's order to put weight on the Irish gelding by feeding him two bushels of oats and a peck of beans a week. As this was the allowance given to the largest coach horses, Cossens thought it excessive. In general, adult saddle horses on the Wimpole estate received between a bushel of oats and six pecks of oats and a peck of beans.[87]

Piecing together disparate snippets of information, Harley's annual stud bill in the early 1720s approached £400.[88] Narrowly employing the formula to the cost of keeping the Levinz Colt for the six years from his birth to January 1729 when the marquis of Carmarthen bid 100 guineas for him, Harley would have spent about £40 on fodder and a proportion of the estimated annual wage bill of £86 for work in the stud. Harley would therefore have made a profit on him. However, if Harley had not bred the colt himself, he would have had to pay out an equivalent sum to acquire a horse of the same quality. Of course, he might have taken the money and groomed another colt to serve as his personal mount. This would have been more problematical, given the failure of the stud to guarantee the production of suitable animals. Anyway, by then Harley was running down the stud. Presumably, the point of a stud was to produce sought-for horses in order to save the money expended in buying them. As Sir Richard Newdigate of Arbury Hall (Warwicks) declared in 1691, 'Wee breed to save Buying'.[89] The Levinz Colt (eventually) was a success, but only a small proportion of the foals produced at Welbeck made the grade, that is, as strong, pacey pads, fit for Harley's service or for sale to his peers. Some remained on the estate, serving as saddle mounts for the staff, and others were disposed of, and rarely at a profit. Not only did Harley have to spend money buying in horses he should have bred at the stud, he made a loss on most of those he sold off. Of the seven recorded sales, only one made a profit, resulting in a deficit of about 200 guineas. The actual shortfall was undoubtedly greater but the fate of most of the foals produced is unknown.[90]

VIII

The vast archive of Harley correspondence not only provides valuable insights into the running of an elite stud in the early eighteenth century but also enables us to look at the horses that were bred and reared there as individuals. As exemplified by the details of the Levinz Colt's life, we know their parentage and can chart the milestones in their lives. We are even aware of the temperament of some of the horses, including that of the Levinz Colt. As the documentation reveals their owner's temperament too, it is possible to infer the nature of his relationship with his horses. In fact, Harley was a distant but demanding figure, leaving the day-to-day running of the stud to his officers but insisting on being kept informed on what was going on. On 22 July 1722, he castigated Hobart for not sending him details of the horses in the stud, which his agent had already done in a mislaid letter.[91] Harley was irascible but also kind-hearted, ending his rebuke with the words, 'when you know me better you will then be sensible that this Letter proceeds from my good Will to you'. And he meant it. His relationship with the Levinz Colt only began when the horse was five years old, but he was aware of his qualities from Hobart's favourable reports. He knew that the colt was very strong and finely proportioned, which made him suitable as a personal mount. He even turned down an offer of 100 guineas in spite of doubts about his temperament. For Harley, the Levinz Colt proved to be a success, an exemplar of the sort of horse he aimed to breed at the stud. One would like to think that horse and rider established a good working relationship, but the horse disappears from the record at this point.

The correspondence also reveals that horses like the Levinz Colt could exercise a degree of agency, influencing their environment directly through their actions or indirectly through what they represented. The differing fortunes of the Levinz Colt and the Careless Colt illustrate the point, the one becoming Harley's personal mount and the other suffering the indignity of castration and rejection. That could have been the Levinz Colt's fate but, to anthropomorphize the animal, he seems to have matured and settled down in the nick of time. Horses could also show their independence by jumping fences and breaking free, by resisting attempts to be broken in and trained, or by showing aggression towards their rider or to other horses. In January 1729, Hobart might have considered that the Levinz Colt was vicious, possibly because he had kicked or bitten another colt a couple of years earlier and this made him question his suitability as Harley's personal mount. The 'rampant' Careless Colt had to be kept apart from the herd. Even accidents had an impact on the stud. In a letter of 2 May 1724, Hobart informed Harley of a fall suffered by Bowron, who had broken his leg when his horse stumbled on a journey. It affected his ability to do his job, for on 11 July he could only walk with two sticks, and, although he could ride, Hobart thought that it would take a few more weeks before he fully recovered.[92]

As possession of fine horses reflected the wealth and status of their elite owners, it was in the owner's interest to treat them well. Compared with the horses belonging to the bulk of the population, they enjoyed a pampered existence. Apart from the stud- and stable-masters, grooms and stable lads, hired specialists broke them in and trained them and treated them when they were ill and injured. They were housed in well-appointed stables (with running water at Welbeck) and the landscape was changed to accommodate them in fenced paddocks. They enjoyed a varied and healthy diet. Of course, in the absence of a 'hands-on' owner like William Cavendish, first duke of Newcastle, much depended upon the attitude of the estate staff whose job it was to tend to them. At Welbeck the supervisory staff, Hobart and Bowron, treated them well and the specialists they hired were efficient and effective. Even so, the establishment was clearly under-staffed by elite standards, which was a constant worry to Hobart. Moreover, as good grooms were hard to find he often had to make do with incompetent or lazy ones. Even in the best-run studs, a horse's fate was contingent on matching the needs of the owner, whether in terms of function or appearance. As many of the horses bred at Welbeck Stud did not meet Harley's requirements, they were sold off if not ridden by an estate official. Similarly, those who were not finely proportioned, were prone to illness or ailment, had a poor action, or possessed a suspect temperament would be rejected. At one time or another, the Levinz Colt failed on three counts but was saved by his strength and appearance (and his sudden maturation). With age, horses lost those attributes that made them desirable in the first place and were moved on, being sold down the social scale or sent to the knacker's yard. As Harley maintained packs of hounds at Welbeck and Wimpole until 1727, it is likely that a number of his superannuated horses ended up as dog meat.

Notes

1 Peter Edwards, *The Horse Trade of Tudor and Stuart England* (Cambridge: Cambridge University Press, 1988) and id., *Horse and Man in Early Modern England* (London: Hambledon-Continuum, 2007).
2 Erica Fudge, *Perceiving Animals: Humans and Beasts in Early Modern English Culture* (Basingstoke: Macmillan, 2000), 2.
3 Erica Fudge, 'Animal Lives', *History Today*, 54/10 (2004), 21–26.
4 Peter Edwards, 'The Decline of an Aristocratic Stud: The Stud of Edward Lord Harley, 2nd Earl of Oxford and Mortimer, at Welbeck (Nottinghamshire), 1717–29', *Economic History Review*, 69 (2016), 870–92. Harley's wife was William Cavendish's great grand-daughter.
5 British Library (hereafter BL), Add. MS 70385, fo. 53.
6 BL, Add. MS 70386, fo. 89. I am grateful to Ann Hyland for confirming Arabians' propensity to change their coat colour.
7 BL, Add. MSS 70385-86; Nottingham University Library, Manuscripts Department, Portland Collection (hereafter NUL), Pl/C/1/1-787.
8 Nottinghamshire Archives Office, Portland MSS [hereafter NAO], DD.5P.1.1. See also Edwards, 'Decline'.
9 Conrad von Heresbach, *Rei rusticae libri quatuor* (Cologne, 1570), translated by Barnaby Googe in *Foure Bookes of Husbandry* (London, 1577), 111r-v; John Lawrence, *A Philosophical and Practical Treatise on Horses, and on the Moral Duties of Man towards the Brute Creation*, 2 vols (London, 1796), i. 78.
10 BL, Add. MS. 70143, fo. 179v.
11 Margaret Cavendish, *The Life of the Thrice Noble, High and Puissant Prince William Cavendishe, Duke, Marquess and Earl of Newcastle* (London, 1667), 67.
12 Edwards, *Horse and Man*, 24.
13 William Cavendish, *A General System of Horsemanship* (London: J.A. Allen, 2000), 12. This is a facsimile reproduction of John Brindley's 1743 English translation of William Cavendish's *La Méthode Nouvelle et Invention Extraordinaire de Dresser les Chevaux* (Antwerp, 1658).
14 William Cavendish, *A New Method, and Extraordinary Invention, to Dress Horses* (London, 1667), sig. b2r.
15 BL, Add. MS 70385, fo. 37.
16 BL, Add. MS 70385, fo. 77.
17 NUL, Pw2/338, 336; Charles Matthew Prior, *Early Records of the Thoroughbred Horse* (London: The Sportsman Office: 1924), 124–25.

18 David Wilkinson, *Early Horse Racing in Yorkshire and the Origins of the Thoroughbred* (York: Old Bald Peg Publications, 2003), 17–18; 'English Foundation Mares', *Thoroughbred Heritage,* www.tbheritage.com/HistoricDams accessed 2 Oct. 2015; Alexander Mackay-Smith, *Speed and the Thoroughbred* (Lanham, MD: Derrydale, 2000), 8, 17, 85–94.

19 Donna Landry, 'The Bloody Shouldered Arabian and Early Modern English Culture', *Criticism*, 46 (2004), 41–69.

20 BL, Add. MS 70143, fos. 287r, 305r.

21 Edwards, *Horse and Man*, 113–14; Daniel Defoe, *A Tour through the Whole Island of Great Britain,* ed. P. Rogers (Harmondsworth: Penguin, 1971), 400.

22 BL, Add. MS 70385, fo. 34.

23 Edwards, *Horse and Man*, 123–24.

24 Edwards, 'Decline'.

25 Edwards, *Horse and Man,* 9, 14, 16; Edwards, *Horse Trade*, 774–75.

26 For the Dun Arabian, see BL, Add. MS 70375, fos. 40, 42, 47. For the Bloody Shouldered Arabian, see BL, Add. MSS 70385, fos. 57, 62 and 70374, fos. 87r, 89r; NUL, Pl/C/1/646; Prior, *Thoroughbred*, 139.

27 BL, Add. MS 70385, fo. 35.

28 Edwards, *Horse and Man*, 24–25.

29 BL, Add. MSS 70388, 26 Mar. 1722 and 70385, fo. 62.

30 Edwards, *Horse and Man*, 25–27.

31 NUL, Pw2/335. A.J. Hibbard, a member of the Old Bald Peg webgroup, which researches the bloodlines of early thoroughbreds, identifies the Whitefoot Mare as the light bay filly foal with a star on her forehead and white above the fetlock of her offside rear foot, got by the Great [Paget] Arab out of a Royal Mare on 29 March 1712.

32 Gervase Markham, *Cauelarice* (London, 1607), II, 2.

33 William Gibson, *The farrier's new guide* (London, 1722), 4.

34 BL, Add. MS 70385, fos. 91, 97, 101, 136.

35 Ibid., fos. 136, 138; NAO, DD.P5/1/1.

36 Edwards, 'Decline'.

37 Ibid. The Whitefoot Mare died of the farcy, an infection of lymphatic system in February 1728. See BL, Add. MS 70386, fo. 42.

38 Edwards, *Horse and Man*, 39–41; Heresbach, *Foure Bookes,* 118v; Gervase Markham, *Markham's Maister-Peece* (London, 1615), I, 222.

39 BL, Add. MS 70385, fo. 37.

40 Ibid., fos. 35, 37, 75, 103.

41 Ibid., fos. 35, 53, 91, 97, 101, 136.

42 A horse whose function was to test the receptivity of the mare.

43 BL, Add. MS 70385, fo. 59.

44 Edwards, *Horse and Man*, 52–67.

45 BL, Add. MSS 70385, fos. 12, 29, 35, 37, 57, 61, 75, 91, 103, 109, 124, 148, 156, 162, 173, 179, 211, 214; 70386, fos. 1, 42, 81, 126.

46 BL, Add. MS 70386, fo. 29; Pl/C/1/591.

47 Louise Curth, *'A Plaine and Easie Waie to Remedie a Horse': Equine Medicine in Early Modern England* (Leiden, Brill: 2013), 209–41; Louise Curth, '"The most excellent of animal creatures", Health Care for Horses in Early Modern England', in Peter Edwards and Elspeth Graham (eds), *The Horse as Cultural Icon* (Leiden: Brill, 2012), 231-7.

48 BL, Add. MSS 70385, fos. 29, 35, 37-8, 148, 150, 156, 162, 171, 173, 179, 195, 197; 70461; 70463; NUL, Pl/C/1/39394, 484, 488, 592, 760.

49 NUL, Pl/C/1/488; BL, Add. MS 70385, fos. 179, 183, 187, 195, 197. Hobart thought that the colt had not recovered from the journey from Wimpole, which suggests lameness, as does the treatment.

50 NAO, DD.5P.1.1, 29 Sep. 1726; Curth, '*A Plaine and Easie Waie*', 167.

51 Edwards, *Horse and Man*, 35–36.

52 Commonly known as riders but called breakers here to avoid confusion.

53 BL, Add. MS 70385, fo. 201.

54 Edwards, *Horse and Man*, 42; John Halfpenny, *The Gentleman's Jockey and Approved Farrier* (London, 1674), 101; Thomas Blundeville, *The Fower Chiefyst Offices belonging to Horsemanship* (London, 1565), 29v–30r; Nicholas Morgan, *The Perfection of Horsemanship* (London, 1609), 98; Cavendish, *A New Method*, 95; Markham, *Cauelarice*, I, 73–74; Michael Baret, *An Hipponomie or the Vineyard of Horsemanship* (London, 1618), 48–49.

55 Gervase Markham, *Countrey Contentments* (London, 1615), II, 37–39.

56 BL, Add. MS 70384, 30 May 1726.

57 NUL, Portland MS Pl/C/1/612.

58 BL, Add. MS 70375, fos. 201, 214.

59 BL, Add. MS 70385, fos. 195, 211; NAO DD.5P/1/1, 18 Nov. 1726, 21 May 1727.

60 BL, Add. MS 70386, fos. 85, 104, 110.

61 It should also be noted that the term 'rude' was used to communicate 'savage' and 'uncivilized', and was therefore redolent of the vocabulary used to describe 'primitive' peoples that so fascinated the Georgian public. To be too rude to be broken also interestingly suggests that, according to Hobart, horses required a certain level of civilization in order to be broken in.

62 BL, Add. MS 70385, fos. 53, 109, 117.

63 Edwards, *Horse and Man*, 74; Richard Blome, *The Gentleman's Recreation* (London, 1686), II, 9; William Durant Cooper (ed.), *Savile Correspondence: Letters to and from Henry Savile* (London: Camden Society, 1858), 160.

64 BL, Add. MS 70398, 7 Aug. 1730.

65 Edwards, 'Decline'.

66 BL, Add. MS 70385, fos. 63, 71.

67 Ibid., fo. 201.

68 Ibid., fo. 175.

69 BL, Add. MS 70398, fo. 2 (16 Nov. 1731).

70 BL, Add. MS 70385, fos. 109, 213–14.

71 BL, Add. MS 70386, fos. 77, 85.

72 BL, Add. MSS 70385, fo. 203; 70386, fo. 81.

73 BL, Add. MS 70386, fos. 77, 83.

74 Ibid., fos. 110, 114, 122; NAO, DD.5P.1.1, 27 Nov. 1729.

75 BL, Add. MS 70386, fo. 89.

76 Ibid., fo. 126. A chafing martingale would be an ill-fitting strap that controlled the horse's head carriage.

77 NAO, DD.5P/1/1, 12 June 1733, 21 Aug. 1734.

78 Susan Pearson and Mary Weismantel, 'Does "The Animal" Exist? Towards a Theory of Social Life with Animals', in Dorothee Brantz (ed.), *Beastly Natures: Animals, Humans, and the Study of History* (Charlottesville: University of Virginia Press, 2010), 26–27.

79 BL, Add. MS 70390, fo. 30 (30 Sep. 1718); NUL Pl/C/1/1/284.

80 BL, Add. MS 70385, fos. 5, 101; NAO, DD.5P/1/1, 14 Sep. 1725. See Edwards, 'Decline' for an overview.

81 BL, Add. MS 70386, fo. 97.

82 Norfolk Record Office, microfilm reel 219/1.

83 BL, Add. MS 70385, fo. 211.

84 NAO, DD.5P.1.1, 4-5 Nov. 1730.

85 BL, Add. MS 70385, fo. 72.

86 NUL, Pw2/329.

87 NUL, Pl/C/1/657.

88   Edwards, 'Decline'.
89   Warwickshire Record Office, Newdigate Collection, CR 136/V/142.
90   Edwards, 'Decline'.
91   BL, Add. MS 70385, fo. 34.
92   Ibid., fos. 109, 115, 118.

Bibliography

Baret, Michael, *An Hipponomie or the Vineyard of Horsemanship* (London, 1618).

Blome, Richard, *The Gentleman's Recreation* (London, 1686).

Brantz, Dorothee (ed.), *Beastly Natures: Animals, Humans, and the Study of History* (Charlottesville: University of Virginia Press, 2010).

Cavendish, Margaret, *The Life of the Thrice Noble, High and Puissant Prince William Cavendishe, Duke, Marquess and Earl of Newcastle* (London, 1667).

Cavendish, William, *La Méthode Nouvelle et Invention Extraordinaire de Dresser les Chevaux* (Antwerp, 1658).

———, *A New Method, and Extraordinary Invention, to Dress Horses* (London, 1667).

———, William, *A General System of Horsemanship*, tr. John Brindley (London: J.A. Allen, 2000).

Cooper, William Durrant (ed.), *Savile correspondence. Letters to and from Henry Savile* (London: Camden Society, 1858).

Defoe, Daniel, *A Tour through the Whole Island of Great Britain*, ed. P. Rogers (Harmondsworth: Penguin, 1971).

Edwards, Peter, *Horse and Man in Early Modern England* (London: Hambledon-Continuum, 2007).

———, 'The Decline of an Aristocratic Stud: Edward, Lord Harley's Stud at Welbeck (Nottinghamshire), 1717–29', *Economic History Review*, 69 (2016), 870–92.

Fudge, *Perceiving Animals: Humans and Beasts in Early Modern English Culture* (Basingstoke: Macmillan, 2000).

———, Erica, 'Animal Lives', *History Today*, 54/10 (2004), 21–6.

Googe, Barnaby, *Foure Books of Husbandry* (London, 1577).

Heresbach, Conrad von, *Rei rusticae libri quatuor* (Cologne, 1570).

Landry, Donna, 'The Bloody Shouldered Arabian and Early Modern English Culture', *Criticism*, 46 (2004), 41–69.

Lawrence, John, *A Philosophical and Practical Treatise on Horses, and on the Moral Duties of Man towards the Brute Creation* (London, 1796).

Mackay-Smith, Alexander, *Speed and the Thoroughbred* (Lanham, MD: Derrydale, 2000).

Markham, Gervase, *Cauelarice* (London, 1607).

———, *Countrey Contentments* (London, 1615).

Pearson, Susan and Mary Weismantel, 'Does "The Animal" Exist? Towards a Theory of Social Life with Animals', in Dorothee Brantz (ed.), *Beastly Natures: Animals, Humans, and the Study of History* (Charlottesville: University of Virginia Press, 2010), 17–37.

Prior, Charles Matthew, *Early Records of the Thoroughbred Horse* (London: The Sportsman Office: 1924).

Wilkinson, David, *Early Horse Racing in Yorkshire and the Origins of the Thoroughbred* (York: Old Bald Peg Publications, 2003).

5   Animals at the table

Performing meat in early modern England and Europe

Karen Raber

John Wecker's *Secrets of Nature* (Latin 1582, English translation by R. Read 1660) offers a recipe for roasting a goose alive. Advising the application of a ring of fire to some 'lively Creature', the recipe includes pots of water to slake the dying goose's thirst, while it 'fl[ies] here and there' within the fire-ring. The cook should baste the goose's head and heart so that 'her inward parts' will roast before she dies: 'when you see her giddy with running, and begin to stumble, her heart wants moisture: she is Rosted, take her up, and set her upon the Table to your Guests, and as you cut her up she will cry continually, that she will be almost all eaten before she be dead.'[1] Thus far, recipes like this one have attracted a limited range of scholarly analysis. Wecker's recipe appears, for instance, in the introduction to Patricia Fumerton and Simon Hunt's collection *Renaissance Culture and the Everyday* (1999), where it serves as a reminder of the casual cruelty of Renaissance practices, which in turn estranges everyday early modern culture for the generation of historicist critics writing in the 1980s and 1990s.[2] In her account of the abattoir that was the early modern kitchen, critic Wendy Wall describes scenes of far more bloody violence accompanying the preparation of meals, drawing out from these consequences for the gendering of household labour.[3] Culinary historians might situate the recipe as an example of the new interest in food's aesthetic complexity during the Renaissance. To animal lovers and vegetarians, the recipe would speak for itself, highlighting the intolerable suffering of living creatures rendered as mere meat for the table: recent work by post-humanist scholars such as Simon Estok and Erica Fudge has discussed early modern resistance to, and rare embrace of, vegetarianism based on the dehumanizing influence of meat-eating exemplified by extreme cases of torture like Wecker's recipe.[4]

Wecker's goose is no lonely outlier. Fumerton's account mentions other examples of such kitchen barbarity as a pig whipped to death or a capon 'pulled' and gutted while alive as evidence that the goose's fate is a common one in early modern cookery. Eels, apparently, were regularly eaten while alive, their motion part of the experience of consumption. A recipe in *The Vivendier* offers a comic take on the goose's lyric performance:

> Get a chicken or any other bird you want, and pluck it alive cleanly in
> hot water. Then get the yolks of 2 or 3 eggs; they should be beaten with

powdered saffron and wheat flour, and distempered with fat broth or with the grease that drips under a roast into the dripping pan. By means of a feather glaze and paint your pullet carefully with this mixture so that its colour looks like roast meat. With this done, and when it is about to be served to the table, put the chicken's head under its wing, and turn it in your hands, rotating it until it is fast asleep. Then set it down on your platter with the other roast meat. When it is about to be carved it will wake up and make off down the table upsetting jugs, goblets and whatnot.[5]

What happens to the naked chicken after it amuses the guests is not reported.

Like many elaborate banquet dishes, Wecker's goose and the *Vivendier's* chicken accomplish a number of things at once. They confuse the distinction between living and dead, between animal and meat; they also collapse the meal's function as sustenance with its function as entertainment. The latter is not surprising since the basic job of a banquet or feast for guests was precisely to affirm or create social ties through a ritualized communal event that often featured grandiose dishes, meant more to impress the diner with the host's wealth or status than to satisfy the palate. These were often interspersed with theatrical, musical or other forms of diversion also nicely calculated to demonstrate the host's authority, good taste, education and other virtues. Both recipes belong to a long tradition of theatricalized, artistically complex meats served at banquet tables, a tradition that extends back at least to the classical literary example of Trimalchio's feast from Petronius's *Satyricon*. That tradition arguably reached its height during the late Renaissance, which often consciously imitated the excesses of classical feasting.

What the goose and chicken recipes do, then, is ensure that the host will be remembered for providing a most entertaining dish, one that results in a miraculous performance by the main course itself. But in early modern Europe, changing habits with regard to meat-eating required the animal at the centre of this performance to take on new roles, new functions. In our own historical moment meat rules the table, unquestioned monarch of the meal, surrounded by fawning courtiers (vegetables and other side-dishes), often enthroned and crowned (resting on beds of starches or doused with sauces). Recent adventures in pink slime and petri dish meats have brought home how hard it is to decentre 'real' meat from this sovereign position: petri dish meat in particular offends through its very status as simulacrum.[6] But it has not always been this way: only at a fairly late date in its etymology, right at the moment Wecker's goose and the *Vivendier's* comical chicken have their moment, did the term 'meat' begin to signify specifically the flesh of a dead animal meant for human consumption in a meal. Prior to the fifteenth century, meat was almost uniformly used as a generic term for all food. The alternative to the current association of 'meat' with cooked animal is the more obsolete use of the word 'flesh', but in the Middle Ages this word referred as often to human beings as to animal bodies and so did not restrictively designate a component of a meal.[7] The etymologies of 'meat' and 'flesh' thus suggest that something was happening culturally in the early modern era that required the role of dead animals at the table to be recoded, to be divided off from other categories of food and bodies.

There are a number of practical reasons why such a transition might have occurred: on the one hand, the huge medieval appetite for meat was displaced during a subsequent period of agricultural change that saw food animals reduced in number, thus making meat consumption more intensely identifiable as a resource for the wealthy, associating it with class and wealth.[8] That shift is reflected in the origin of terms for animals themselves, as opposed to the meats they become: 'sheep' and 'pig', for instance, are both derived from the rather more common Old English, while 'mutton' and 'pork' are Old French in origin, attesting perhaps to the more patrician nature of animals' culinary identity. In their survey of French cookery books, Philip and Mary Hyman observe that the very popular *Le Thresor de Santé* (1607) devotes fully one half of its length to recipes for meat.[9] When the advent of more successful and productive agriculture allowed more people access to meat at meals, meat thus gained a tremendous cultural cachet. A growing role for the culinary arts in ever broader segments of society throughout the sixteenth and seventeenth centuries also focused the attention of many (even those who still could not afford much) on feats of cookery applied to meat; meanwhile, widening popular concern for the medical role of meat in dietary regimes encouraged people to think carefully about distinctions among meats and between meat and other foods. Whatever the economic, scientific or other practical reasons for meat's changing role, it was also transformed into a cultural focal point through its various representations as an object, one engaged in complex interactions with human bodies, with other meats, with other 'players' at the banquet table. But that new status for meat only makes more potent the problem of establishing what it is that 'makes meat'. Is the living animal already incipient meat?[10] At what point in its metamorphosis does an animal become meat: when slaughtered, when divided by the butcher, when cooked, when eaten? 'In the eating encounter, all bodies are shown to be but temporary congealments of a materiality that is a process of becoming, is hustle and flow punctuated by sedimentation and substance.'[11] Before being consumed, meat acts on human senses and imagination: odour, texture, and taste all simultaneously generate autonomic responses in human anatomy via the human brain (salivating, for instance), as well as imaginative, memory-driven or aesthetic appreciation in the individual. After consumption, flesh melts into flesh, becomes categorically indivisible, yet can generate discomfort, illness in the short term or obesity and debility in the long term. The simplistic observation that 'you are what you eat' hides a rich and complex set of processes and intra-actions that this essay will probe.[12]

As I noted above, most discussions of early modern meat are found in works of culinary history, animal studies or activist anti-meat criticism. While welcome and a clear inspiration to this project, the various agendas of such recent work have tended to ignore or erase the nuanced process by which cultural dominance has accorded meat as a main part of meals, and the consequent cultural negotiations of meat's inherent complexity as a performer at the table. In this study, I focus narrowly on meat's position as an actor in the theatre of the meal. In the interests of understanding how (usually mostly) dead matter can be said to perform, I mobilize the work of the 'new materialists,' particularly that of Jane Bennett, who offer a way to talk about the metaphors mobilized by and through meats, about meat's role as actant, and about what Bennett calls its 'vagabond' quality, and its vitality-in-death.[13] In what follows, I will take up the question of what is at stake in the appearance of two groups of performing meats included in early modern feasts and banquets: first, zombie meats, in the vein of Wecker's goose and the *Vivendier*'s chicken; second, the related creation of early modern 'transgenic' or 'masquerading' meats through the technique of engastration (the stuffing of one meat with another or several others) and engastration's close analogues, 're-dressed' meats – meats made up to look like living versions of themselves, or like other creatures or objects. What these 'performing meats' have in common is the multidimensionality of their required signification. They do not simply entertain, although certainly that is part of their purpose. Rather I argue their performances at the dinner table illuminate early modern ideas about, and desires regarding, the living animals that provided a meaningful dietary mainstay. I will not contend that early moderns in any way were appalled or revolted by the collapse of theatrical performance and the rendering of animals as meat – that is, I will not suggest that Wecker's or any other such recipe aroused ethical concerns about animals themselves. I argue instead that early modern banquets created performed and performing meats that violated species and other categories, and that while this theatre of meat announced and celebrated human exceptionalism and human control over nature by testifying to the creative and transformative power of the human cook (and host), it also revealed the limits of that power by conceding or granting to animal flesh a type of agency – what Judith Butler might call the 'performativity' of meat – in the process of making it act out a part in a meal.[14] Ultimately, I maintain that what meat 'performed' was itself – all the distortions, complications and ideological dimensions of its production *as meat*.

Zombie meats

Whether or not Wecker's description of the recipe for cooking a goose alive is meant to be a real recipe is debatable. For one thing, it is included in a book that calls itself 'The Secrets of Nature', suggesting that it belongs to the genre of alchemical or occult literature; the volume shares some aspects with natural philosophies, including, for example, remedies for various diseases and offering a compendious and inclusive treatment of the various creatures it discusses alongside recipes and other kitchen lore. Wecker's may not be the only recipe of its kind in early modern cookbooks, but no corresponding descriptions of a 'live' animal consumed at table during an actual banquet seem to survive (apart from those involving eels) – not to mention the fact that it is not clear if the recipe could even work reliably. Yet if it does not represent a material practice regarding meat in the past, it certainly represents a *fantasy* about what cooking and eating can achieve. The questions I address in this section are: what are that fantasy and what are its implications? What is the appeal of Wecker's zombie goose? What role, exactly, is it performing for his readers, and to what ends?

The theatricality of early modern banqueting has been amply charted by historians and literary-cultural scholars. Ken Albala remarks about the theatre of cooking and eating that 'any meal... contains a script. It might be said that every participant in the eating event is equally an actor.'[15] Roy Strong describes the elaboration of food items in late medieval and early modern banquets, noting that 'eating [had] become an element in a vast theatrical production.' At one feast given by Gaston IV at Tours in 1457, a series of entremets divided numerous courses; 'mobile scenery, costumed actors, singers, musicians and dancers' enlivened the scene, along with a culinary 'finale' involving a 'heraldic menageries sculpted in sugar: lions, stags, monkeys, and various other birds and beasts, each holding in paw or beak the arms of the Hungarian king' who was being honoured.[16] As medieval appetites for both food and spectacle evolved, cookery books and famous cooks ushered in a new era of cuisine, influenced by many factors, including the use of spices, hints of Arab cuisine, new culinary techniques, and so on. One consequence of this expanding importance and investment in food was a new standard for aspects like 'form and colour' – Strong remarks that 'the physical side of eating is being displaced by the aesthetic pleasure of looking.'[17] Not displaced, perhaps, but certainly melded with the idea of food as both visually and gustatorily consumable. In 1587, *Epulario, or the Italian Banquet*, translated from Italian for English cooks, gives a description of a type of dish now familiar to us only because of its appearance in a nursery rhyme: 'To make Pies that the Birds may be alive in them, and flie out when it is cut up'.[18] Inside a larger 'coffin' or shell, the recipe calls for the insertion of a smaller, genuinely edible pie that is surrounded with 'small live birds' that will fly out when the larger shell is opened, 'to the delight and pleasure shew to the company.' The emphasis on 'shewing' is recognized a moment later when the recipe advises the smaller pie is there for actual eating 'that they [the guests] shall not bee altogether mocked.' To be entertained by the performance of the birds evokes pleasure; for the visual spectacle to be all, would be to mock the diners' appetites for consumption of another kind. Gaze and palate must both be satisfied. Strong describes castles created out of meats, sugars, flour and other ingredients populated by creatures, among them a full-grown stag, boar, rabbits and so on, the latter clearly 'meant to be eaten,' a boar's head breathing fire, fountains of wine, enormous confections large enough to contain human musicians.[19] The crying of Wecker's goose is a much-redacted variant of such a grandiose display, blending gustatory and auditory pleasure when the presumably faint sounds of the goose's voice accompany consumption of her tasty flesh.[20] From the fourteenth to the seventeenth century, food functioned much as it had always done, as propaganda and display, but with increasing attention to its potential as art, sculpture and even full-bore theatre.[21]

I would isolate two aspects of the recipe to consider more closely: the theatrical performance of the goose's actual death, and its performance for the diner once served. Wecker presumably describes an act of cookery that happens in a kitchen well away from the guests who will partake of the dish, yet he does so in excruciating detail, constructing a scene that resembles nothing so much as a miniature drama. Surrounded by kitchen staff, including the cook, who must bank the fires that roast her, the goose has an audience to her immediate suffering, mirrored in the reading audience of the cookbook once the recipe is printed. She is active, flying around looking for escape, periodically basted with water to encourage her further struggles. A death scene more lingering and pathetic could hardly be found on the early modern stage, suggesting that what matters in this recipe is not only the eventual dish that results, but the pleasure (whatever that consists of) in vicariously witnessing this transition from living, 'lively' animation to zombie-like living death. That pleasure is so great, so compelling, that readers who might never experience it directly are being invited, through the printing of the recipe, to mentally re-enact it. What do Wecker's readers gain from this possibly spurious description? A moment of insight into how life can be preserved in the face of even the most deadly treatment – insight, that is, into the very moment that is obscured to the most educated early modern reader, even to medical experts, in a world where the fact of death was everywhere, but knowledge of what precisely it *was*, nowhere. Or do they enjoy the pleasure of witnessing an act of torture that results, because of its repressed and controlled violence, in a moment of beauty (the song as she is eaten)? If this is the consequence, the scene then evokes a tradition of torture and martyrdom that makes the goose a Christ figure, sacrificed in a parody (or is it?) of the Passion and the Eucharist.[22] Is this latter possibility the reason that the goose's consumption, and its textual reproduction, are different – one involves partaking of the flesh, the other an imaginative investment in the scene of death itself? If so, what does that make the recipe book? It is hard not to think here of the current craze for 'food porn', which turns images or representations of recipes into a substitute for sexual pleasure. But in this case, the 'sexual' is something else, something closer to a transformative embodied experience that conjures sensory experiences usually absent from ordinary meals.

Wecker's goose had, of course, already been the target of human transformation on a global scale. Geese were both wild and domesticated creatures in early modern Europe, of course. However, because Wecker's recipe comes amidst a long section on domesticated fowl of all kinds, including ducks, chickens, and pigeons, we can be reasonably sure this is the domestic variety. Wecker also advises that the goose is only the preferred fowl for this dish, but also names ducks or 'some lively creature' as options. Geese were domesticated as early as 3000 BCE and are ubiquitous rather than limited to certain regions. In addition to providing eggs and flesh for the table, geese are good watchfowl, especially in groups, since they are both extremely loud and potentially aggressive toward other animals or unfamiliar humans. For these reasons, the goose should be considered a touchstone for the entire concept and process of domestication. And domestication is human improvement in its most concrete and generalized form. Before arriving at the dinner table, then, a goose is already a 'redressed' creature, one whose physiology and behaviour have been meddled with by human agency, and have been turned into a mirror for human power over nature. Albala observes that the predominance of domesticated over wild meats grew exponentially in the late sixteenth and early seventeenth centuries, and that the preference for meats generated from human control over nature, instead of those derived directly from nature itself, had moral and cultural connotations, so that wild meats came to be associated with unrefined or 'wild' characteristics in those who ate them. Instead of noble hunters, early modern foodies were beginning to think of themselves equally as scholars and collectors who appreciated the finer, lighter meats.[23]

This goose, then, is being mobilized as a performer in something more complex than the mere display of wealth and privilege: her 'speech,' her faint movements, even the knowledge of the means of her preparation enhance the aesthetic dimension of the meal itself, while the knowledge of, or report of the method of her preparation might fashion a new kind of textual experience once translated to print. But it is worth thinking about what is different when a living animal is thus made actor in the drama of eating. In addition to the animal being made compliant through a prior process of domestication, the bird dying here is produced as the sauce to her own demise. While we might think this a horrific thing, early moderns more likely would have appreciated the ideological content of the moment – an animal serenading a diner whose consumption endorses the act of eating as an aestheticized assertion of human control over creation, and therefore also over death. At the same time, however, the 'cut' that should mark the goose's flesh as object, as dead and therefore edible meat, instead disrupts any neat distinction:[24] the diner carves into an animal that announces by voice and gesture that it is still animate, conscious, a participant in the drama of the table. The bloody drama usually assumed to end in the kitchen arrives at the table; instead of passive audience, guests become actors on stage themselves completing the act of killing Wecker's goose and so fixing it as meat. If the difference between the raw and the cooked is culture, then this goose arrives only slightly cooked. As to the cook and guest who rely on a meal of meat as a demonstration of exclusively human agency: their goose is cooked indeed when they enlist this goose in their performance of her death.

Make your own (dead) animal

Everywhere in the early modern kitchen, an observer could find examples of transmutation, things being turned into other things, often involving various forms of meat. 'Turn your meat,' writes Lady Elinor Fettiplace in one recipe, 'to pure blood.'[25] Wendy Wall notes that cookbooks 'underscore the importance of flesh mutating into flesh... everywhere hearkening toward dinner's vitality and the precariousness of embodiment.'[26] Food was used to create almost anything, from small objects to entire environments: Strong describes fake gardens made of sugar, vessels and instruments, statues and sculptures, even entire buildings made of food.[27] The feast was a 'game of deceit,' with edible trenchers, cups and so on – but also featuring meats layered or fused within, around and on top of other meats, disguised as other creatures or as their own living selves.[28]

Part of the reason for meat's 'vagabond' nature in the early modern world is the degree to which it was not yet cultivated to enhance its more consumable qualities – that is, flesh was hard to eat because generally it was quite tough (and early modern teeth were probably not always reliably strong enough to tear and chew it fully). Thus, the vast majority of recipes in the period involving meat require its dis-integration through extreme cooking techniques. Nearly every meat is stewed, seethed or minced and then mixed, stuffed or sauced with other ingredients; many meat dishes end with the resulting 'paste' reconstituted through baking or incorporation into puddings, hashes, or other blended dishes. Roasted meats are actually fairly rare (usually appearing at larger banquets in the late Middle Ages) and often limited to more tender animals like fowl. What this means is that early modern meat dishes obscure their origins: one could not necessarily perceive in the resulting food the shape or other physical attributes of the living animal. Even the sheer act of butchering transformed food into something vastly different from its first incarnation, while even the simplest styles of cooking were by definition processes of transformation. The preface to John Day's *The Travels of the Three English Brothers* (1607) compares playwriting to cooking by noting the additions, subtractions and obfuscations of both thus: 'Who gives a fowle to his cook to dress / Likewise expects to have a fowle againe, / Though in the Cookes laborious workmanship / Much may be deminisht, som–what added / The losse of feathers and the gaine of sauce.'[29] While a diner might expect her cook to send a 'fowl' back from the kitchen, she accepted that it might arrive in new, saucy attire following the mysterious process of re-creation applied to it.

At the simplest level, by creating re-formed and re-dressed dishes cooks were merely restoring visual cues to the animal's identity and a less ambiguous connection between the transformed meat and its prior condition as a live animal. *Epulario*, for instance, includes a recipe for how 'to dresse a Peacocke with all his feathers' that produces a dish that 'seems to be alive'.[30] The cook removes the bird's feathers and skin, cooks its meat, then restuffs it with its own flesh and re-feathers it. While this is the most frequent process cited in recipes, it turns out that the dis-integration of meat through cooking opened the door to much more inventive results. Rendering meat edible also provided an opportunity to quite literally make meats 'cross-dress,' like one of Shakespeare's boy actors done up in women's garb. Early modern meat thus becomes the material of experiments with nature, transforming and translating what *was* into what *might be*. If 'dressing' (meaning to form, order, arrange, straighten or manage) referred to meat's preparation either for cooking or for serving, then we might say that meats were also 're-dressed' in other attire for their appearance at the dinner or banquet table: that is, they were re-clothed and amended or remedied in the process.

Consider the turducken, a turkey stuffed with a duck, which is in turn stuffed with a chicken, a dish now primarily served at Thanksgiving feasts in the U.S. Although its name is new (dating, according to the OED, only from the 1980s), its origins lie in the period under investigation, in the fascination with engastration that informs many early recipe books and banquet tables. The English call three-bird roasts 'royal roasts,' and they derive from the very old process of boning and then stuffing a portion of meat for added flavour and texture. The Tudor Christmas pie, which dates back to the tables of Henry VIII if not before, placed an engastrated turkey inside a 'coffin' of pastry (probably to allow it to cook well without burning). Turkeys were, Bruce Boehrer notes, objects of 'high-end gastronomic desire,' due to their exotic origins and their association with the peacock, a traditional bird at the most splendid feasts.[31] Of course, they were also large enough birds to allow the enclosure of various other game and domesticated fowl within, a not insignificant part of their appeal. It is tempting to link their hidden-yet-populous interiors with their obscurity to early modern natural histories, since turkeys also presented a challenge to European lineages of species: was the bird entirely new, or a true relative of peacocks (a question with both religious and scientific implications)? At the very least, it is possible to observe that the turkey's indefinite origins, in a fortuitous accident of gastronomy, align nicely with their indefinite identity as highly engastratable birds.

Perhaps the grandeur of the turkey has led us to overlook the smaller animals inside but from the perspective of the chicken in a turducken the subsequent layers are a form of cloaking device, concealing its 'nature' until the moment when the turkey is carved and reveals itself to be not a singular dead animal, but one inhabited by other creatures. It is, thus, a variation on the many surprise theatrical food-based revelations included in famous banquets. Jeffrey Hudson, court dwarf to Henrietta Maria, was first presented to her leaping out of a pie crust. Pope Gregory XIII offered a miniature model of the Castello di St. Angelo out of which came rabbits, partridges and other small creatures.[32] The *Vivendier*'s dormant chicken is one version of these surprises, which incorporated – literally, made part of the corpse – living animals violating the expectation of dead meat, as well as elements of theatre wedded to gastronomic techniques. The chicken at the heart of the turducken continues the experience of novelty and wonder among diners in all these episodes, aided by the turkey's and duck's savoury flavours, but also by their function as disguise.

The turducken is really quite a tame critter: the most extreme example of animal experimentation comes in attempts to create entirely new creatures from dead flesh. For his banquet in honour of the French King, Francis I, at the Field of the Cloth of Gold in 1520, Henry VIII's cooks whipped up a 'cockentrice,' by sewing together the head of a pig and the rear end of a chicken. While it might look like a bizarre violation of nature, it was not so rare a dish, having already graced the table of John Stafford, Bishop of Bath, in 1425 and probably many more banquets besides.[33] If Wecker's goose and the *Viviendier*'s chicken are proto-zombies, then perhaps these 'redressed' meats count as early experiments in transgenesis – the manipulation of animal DNA to produce new species, to recode dead flesh and give it new naturalized-unnatural forms. What cannot be found in nature is produced by human ingenuity and intervention, which amends and improves – that is, redresses simple meats, giving them new identities.

What do engastric or cross-dressed meats in early modern cookery tell us, either about meat, or about what it represents? While the engastration of meats can be assimilated to other forms of transformation at the banquet table (like *Schauessen* or *trionfi* – 'display dishes' or 'triumphs' – confections in all sorts of shapes and forms made out of a variety of materials), making meat into a simulacrum of itself or of other meats suggests that 'meat' functions as figurative and symbolic *matter* – it is in itself metaphor, or perhaps an example of what Ian Bogost calls 'metaphorphosis' in the sense that 'meat' as a descriptor of dead flesh is detached from any 'natural' or confirmed 'thing' in the world and instead becomes a thing in itself.[34] Each of these masquerading dishes is an ontologically confused and confusing thing, its existence made possible precisely because of the mobile (in every sense of this word) matter of a thing (dead flesh) that was once a living object (the animal). The boundary crossings of these masquerading meats can be assimilated to the same narrative as our performing meats above. However, I think the process of creating redressed meats carries a particular cost: by provoking cooks and diners to reconsider meat's inert, passive status and the reliability of meat's self-identity, they raise questions about meat's dangerous potential influence on the identity and status first of the animal that is transformed into meat, and next of the eater who consumes it.

Early modern dietaries and medical texts posited a humoral human body, porous and so vulnerable to external influences, constantly struggling to achieve equilibrium. Geography, class status, gender, and other factors could influence an individual's basic humoral complexion, while everything from air to food could disturb the precarious balance of that body's internal machinery.[35] Whole categories of meats were understood to define the bodies that ate them: pork, for instance, was a lower-class dish, suitable for crude palates and crude bodies, while tender fowl was for more refined diets. Albala notes that Charles Estienne's *de Nutrimentis* (1550) 'excised many foods completely in consideration of health,' including onions and garlic because they were only 'appropriate for barbers and journeymen'.[36] Luigi Cardano found no room for apples in any healthy diet. Food was never simply fuel: it was physic for a range of ailments, with effects on everything from individual morality to national identity. 'All acts of ingestion and excretion,' Michael Schoenfeldt argues, were 'very literal acts of self-fashioning'.[37] Gluttony was a frequent target of criticism in dietaries, followed by drunkenness; but excessive consumption of meat also repeatedly draws opprobrium from dietary authors. William Bullein recommends only certain meats for creating 'good blood,' including many fowl and roasted veal or boiled mutton, while Andrew Boorde praises pheasant as a bird that 'doth comfort the brayne and stomach,' along with other small birds whose flesh is nourishing but does not 'ingender melancholy'.[38] Like many English dietaries, Boorde's applauds beef as a dish for Englishmen, as long as it is neither too old nor from a female animal, either of which cause 'leprous humours'.[39] Pork is an especially vexed subject, even when it is not remade as an early modern proto-cyborg: Fudge recounts the various positions for and against but most agreed that it was risky because of the pig's association with uncleanliness and with cannibalism.[40]

If one cannot tell the identity of the meat that one consumes, however, then obviously any prescription regarding appropriate consumption of the stuff is rendered ineffectual. Moreover, if meats can be recoded, not merely as different meats (as in the case of layered and blended meats) but as completely new creatures (as in the case of the cockentryce), then the entire edifice that rests on the Galenic system falls apart. Instead of policing social, political, national and other boundaries, meat violates the whole notion of decipherable categories. Again, Bennett's use of the term 'vagabond' describes meat's inherent variability, its itinerant nature, resistant to the kind of fixity required by dietary regimes of the period. In an accident of history and language, we might recall here that early modern stage performers, actors in the public theatre, were regarded as vagabonds and 'masterless men' by authorities. Like human performers, banquet meats promised a theatre of order and discrimination, but in their mobility might deliver the opposite.

If transgenic meats expose meat's susceptibility to transformation, and redressed or masquerading meats suggest the difficulty diners might have in even recognizing the meat being served to them, then not only does meat not enable the policing of social, moral, political and other categories as it is supposed to, but as a 'vagabond' might irrevocably lead to the complete collapse of all categories. The anxiety embodied in meat revolves around its mobility – it is always in the process of becoming something else, animal becoming flesh, flesh becoming 'meat', meat being cooked, cooked meats being consumed, consumed meats becoming (human) flesh again, and so on. At each stage, what meat is and is not is uncertain; in the last stages when animal flesh is transmuted into human flesh, meat enacts a mingling of bodies that confronts the diner with the porousness of her body and its essential material instability. Matter is never itself; it is always becoming other. Engastric concoctions like the turducken and cross-species confabulations like the cockentryce, I believe, deliberately reproduce this indistinction as a circumscribed by-product of human intervention in the making of meat, rather than in its most global and threatening form, as definitional of all flesh; they do so precisely in order to confront and defang the anxieties aroused by the very act of eating a dead animal. In light of this possibility, it is useful to consider for a moment the practice of redressing meats as the animals that the flesh originally belonged to. A peacock dressed up in its feathers is dead meat masquerading as living animal – or, in another formulation, an animal masquerading 'as itself,' just in a more culinarily compliant form. Such redressing attempts to introduce stability and a different kind of vitality to the dead, confused and confusing object being presented to diners. But what does it mean to say the bird is dressed as 'itself'? What 'self' does the bird – dead, dismembered, mixed with other ingredients, reassembled, shaped, and re-feathered – have? The act of culinary re-dressing imports a fantasy of self-identity, of a prior subjectivity invested in the living animal. While not precisely a process that recognizes some individuality of a bird, re-dressing it at least implies respect for species identity as a partial substitute. One would not want a guest to cut into an object that did not proclaim itself a specific type of animal.

Early modern meat had to be made, first by cultivation of living animals as domesticated breeds suitable for consumption, then as flayed and dismembered carcasses, then as transformed culinary objects. What makes meat, however, is also its performance of itself (of *a* 'self') throughout the penultimate stage of transformation, the meal, *as meat*. Only then can a guest recognize the various codes that render meat edible and take it on its final journey, to mingle with human flesh from within. Wecker's goose is not meat until after its body is cut into by the diner, extending the 'making' process from kitchen to table. Cross-dressed meats masquerade as living creatures, as themselves in lively form, until they are dismembered a second time by the diner who restores the certainty of death and renewed life as human, not animal, flesh. At the banquet table, 'performance,' in its more common sense of theatrical action, bleeds – literally – into 'performativity,' the construction of matter *as matter*. In its tales of zombie geese and chickens, its transgenic and cross-dressed pheasants, turkeys, and ducks, early modern culinary practice stages the dangerous cultural drama of becoming meat.

Notes

1. The recipe appears on 148. Wecker's is not precisely a cookbook, but rather a grab-bag of 'secrets' in various fields, accompanied by recipes both for foods and for medicines.
2. Patricia Fumerton and Simon Hunt (eds), *Renaissance Culture and the Everyday* (Philadelphia: University of Pennsylvania Press, 1999), 2.
3. Wendy Wall, *Staging Domesticity: Household Work and English Identity in Early Modern Drama* (Cambridge: Cambridge University Press, 2002), 338. The afterlife of Wecker's recipe includes its appearance in M.F.K. Fisher's post-war *How to Cook a Wolf* (New York: Farrar, Straus and Giroux, 1942) – in a chapter called 'How to Make a Pigeon Cry' the title of which actually refers to a quotation from Jonathan Swift – as part of her encouragement to cooks faced with rationing that their talents could conquer and transform the most bizarre kinds of meats.

4 See Erica Fudge, 'Saying Nothing Concerning the Same: On Dominion, Purity, and Meat in Early Modern England,' in id. (ed.), *Renaissance Beasts: Of Animals, Humans and Other Wonderful Creatures* (Urbana: University of Illinois Press, 2004), 70–86; and Simon Estok's 'Theory from the Fringes: Animals, Ecocriticism, Shakespeare,' *Mosaic*, 40 (2007), 61–78. Fudge is concerned with the justifications of early modern resistance versus acceptance of vegetarianism, while Estok argues that Shakespeare's plays sanction vegetarianism in many of their representations of meat-eating.

5 *The Vivendier* (ca.1450), Gesamthochschul-Bibliothek, Kassel, 4° Ms. Med. 1. Quotation is from Terence Scully, *The Vivendier: A Critical Edition with English Translation* (Totnes: Prospect Books, 1997), 81.

6 In August 2013, Mark Post, a vascular biologist, offered his lab-grown meat in a publicity stunt for which it was cooked as a hamburger by a famous chef and tasted by two food critics (see, for instance www.theguardian.com/science/2013/aug/05/first-hamburger-lab-grown-meat-press-conference or www.the-scientist.com/?articles.view/articleNo/36889/title/Lab-Grown-Burger-Taste-Test/, both accessed 8 Sep. 2016, for online articles covering the event). Public reaction ran the gamut, but one constant was the momentary wince at the mere thought of consuming meat that did not have its origins in an authentic cow. Safety and cost concerns aside, if petri dish meat fails it will be this reaction that kills it, I would argue.

7 The OED gives initial instances from 1325 and 1475 for this more narrow usage; until that moment, meat was anything that could be eaten, from vegetables to sweets. The Middle English Dictionary gives numerous versions of 'meat' that refer to all forms of solid food (as opposed to drink, for example); only in one sub-category is 'meat' defined as derived from animals, and then it is indistinguishable from the 'meat' of fish or any other form of flesh. A review of the Middle English Corpus reveals numerous uses of 'meat' for virtually any food, from individual items like onions to concoctions like possets or desserts. Noëllie Vialles notes that the same shift happens to the French *viande*, in *Animal to Edible*, trans. J. A. Underwood (Cambridge: Cambridge University Press, 1994), 4. Meanwhile, 'flesh' functions as a reference to the human body (as in 'all flesh is weak') in the medieval literature, as well as to communion bread, or to the muscle and other tissues of a living mammal (thus exclusive of fish or fowl).

8 See Ken Albala, *The Banquet: Dining in the Great Courts of Late Renaissance Europe* (Chicago: University of Illinois Press, 2007) and Roy Strong, *Feast: A History of Grand Eating* (New York: Harcourt Inc., 2002). In his introduction to early modern food in Jean-Louis Flandrin and Massimo Montanari (eds), *Food: A Culinary History from Antiquity to the Present*, trans. Albert Sonnenfeld (New York: Columbia University Press, 1999), Jean-Louis Flandrin points out that over a long stretch of time from the late Middle Ages through the eighteenth century, archaeological evidence suggests that 'diet ceased to be determined by the hazards of production and began to be shaped instead by consumer preference' (405). This was especially true for the social elites, although Flandrin's cited example is a French monastery.

9 Philip and Mary Hyman, 'Printing the Kitchen, French Cookbooks 1480–1800', in Flandrin and Montanari (eds), *Food: A Culinary History*, 396. Albala observes that because domesticated meats represented the bending of nature to human will, they may have been of increasing value for their symbolic function at the same time (*The Banquet*, 33).

10 Noëllie Vialles points out that requirements for bloodletting as the means for an animal's death in butchering (and proscriptions on using animals that die by disease or accident) and status as a 'permitted species' guarantee that only certain animals, and only living ones, make it to the dinner table. Vailles, *Animal to Edible*, 5.

11 Jane Bennett, *Vibrant Matter: A Political Ecology of Things* (Durham, NC: Duke University Press, 2010), 49.

12 The term 'intra-action' belongs to Karen Barad, as does a version of the concept of 'performativity'. I intend both terms to resonate throughout this essay. In her work Barad joins everyone from Foucault to Haraway to Latour to Butler in embracing the body and resisting what she calls representationalism (the primacy of language to construct realities); however, she addresses what she sees as a gap in these theorists' work, the specific process by which discourse and bodies interact. Barad argues for the body's, and all matter's, agential realism (an account of human and non-human ontology that takes seriously the idea of matter's agency, so that rather than 'words' and 'things' the world consists of relationalities that are material in nature). Her neologism, 'intra-action,' insists that there are no pre-existing entities before relation; that only through intra-action do the boundaries of phenomena come to exist. What is useful about Barad is that she doesn't stop with the connotations of 'performative' as, for example, Butler uses the term, but strenuously resists the tendency to assimilate 'performance' back to divided ontologies of being and/vs. representation, problematizing implicit causality in the way we think about these things. See Karen Barad, 'Posthumanist Performativity: Toward an Understanding of How Matter Comes to Matter,' *Signs*, 28 (2003), 801–31.

13 By 'vagabond,' Bennett means 'a propensity for continuous variation,' *Vibrant Matter,* 50. The term 'actant' is used in Bruno Latour's actor-network theory to describe a mediating object with a degree of agency, but without the subject-position – or the philosophical baggage – associated with the term 'agent'. See Bruno Latour, *Reassembling the Social: An Introduction to Actor Network Theory* (Oxford: Oxford University Press, 2005), 71.

14 Butler originally proposed the idea of gender as performative in *Gender Trouble: Gender and the Subversion of Identity* (New York: Routledge, 1990), although the term and what it signifies have been under constant revision and extension by Butler and others since.

15 Albala, *The Banquet*, 4.

16 Strong, *Feast*, 75.

17 Strong, *Feast*, 85.

18 *Epulario or the Italian Banquet* (London, 1590), sig. B4r.

19 Strong, *Feast*, 118–19.

20 As noted above, in M. F. K. Fisher's cookbook, the pigeon who cries does so for Jonathan Swift: 'Here's a pigeon so finely roasted, it cries come eat me' writes Swift in *Polite Conversation* (1738), as Mr. Neverout tempts Miss Notable at a meal. Fisher has uncovered the attraction in Wecker's crying goose – the animal that audibly invites its own consumption makes a tastier tidbit, saucing itself with the addition of sound, the one sense that isn't usually directly addressed by strictly culinary prodigies. See 'Simon Wagstaff' [Jonathan Swift], *A Complete Collection of Genteel and Ingenious Conversation* (London, 1738), 136.

21 Albala calls it propaganda, reflecting on how food choices could convey volumes about colonialist practices, morality, and national identity. *The Banquet* 5.

22 Nearly every dead animal in early modern cultural representations gets cast at some point as an image of sacrifice and thus tied to the Passion and/or the Eucharist.

23 Albala, *The Banquet*, 33.

24 Barad, 'Posthumanist Performativity', 815, contrasts the Cartesian 'cut,' which relies on inherent differences between subject and object, to an 'agential cut,' that creates a 'local resolution *within* the phenomemon of the inherent ontological indeterminacy.'

25 Elinor Fettiplace, *Elinor Fettiplace's Receipt Book*, ed. Hilary Spurling (London: Faber and Faber, 1986), 334.

26 Wall, *Staging Domesticity*, 338.
27 Strong, *Feast*, 188–97.
28 Wall, *Staging Domesticity*, 335.
29 John Day, *The Trauailes of the Three English Brothers* (London, 1607), sig. A2r.
30 [Giovanne de Rosselli], *Epulario, Or, The Italian Banquet* (London, 1598), sig. C[1]r.
31 Bruce Boehrer, *Animal Characters: Nonhuman Beings in Early Modern Literature* (Philadelphia: University of Pennsylvania Press, 2011), 136–37. Flandrin, in his essay 'Dietary Choices and Culinary Technique, 1500–1800', notes that the turkey was the sole innovation among fowl in the period, and maintained its gastronomic 'status' despite increasing availability (it was easily domesticated). See Flandrin and Montanari (eds), *Food: A Culinary History*, 404.
32 Strong, *Feast*, 181.
33 British Library, Harleian MS 279 (ca. 1430), fo. 62, lists a 'cockyntryche' among the banquet dishes.
34 See Ian Bogost, *Alien Phenomenology or What It's Like to be a Thing* (Minneapolis: University of Minnesota Press, 2012), 66. Bogost coins the term to address how metaphor can function not merely representationally, but as a 'means to apprehend reality.'
35 See Ken Albala, *Eating Right in the Renaissance* (Berkeley: University of California Press, 2002) and Joan Fitzpatrick, *Food in Shakespeare: Early Modern Dietaries and the Plays* (Burlington, VT: Ashgate Press, 2007).
36 Albala, *Eating Right*, 34.
37 Michael Schoenfeldt, 'Fables of the Belly in Early Modern England,' in Carla Mazzio and David Hillman (eds), *The Body In Parts: Fantasies of Corporeality in Early Modern Europe* (New York: Routledge, 1997), 243.
38 William Bullein, *The Government of Health* (London, 1595), 26.
39 Andrew Boorde, *A Compendious Regiment, or Dietarie* (London, 1576), sigs. F1r, F2v.
40 On the idea that eating a pig was always potentially an act of cannibalism (because pigs ate humans along with every other form of food), see Fudge, 'Saying Nothing Concerning the Same.'

Bibliography

Manuscript Sources

British Library, London
   Harleian Manuscripts
   279 Cookery Book, *c.*1430.

Printed Material

Albala, Ken, *The Banquet: Dining in the Great Courts of Late Renaissance Europe* (Chicago: University of Illinois Press, 2007).
———, *Eating Right in the Renaissance* (Berkeley: University of California Press, 2002).
Anon., *Epulario or the Italian Banquet* (London, 1590).
Barad, Karen, 'Posthumanist Performativity: Toward an Understanding of How Matter Comes to Matter', *Signs*, 28 (2003), 801–31.
Bennett, Jane, *Vibrant Matter: A Political Ecology of Things* (Durham, NC: Duke University Press, 2010).
Boehrer, Bruce, *Animal Characters: Nonhuman Beings in Early Modern Literature* (Philadelphia: University of Pennsylvania Press, 2011).

Bogost, Ian, *Alien Phenomenology or What It's Like to Be a Thing* (Minneapolis: University of Minnesota Press, 2012).

Boorde, Andrew, *A Compendious Regiment or a Dyetary* (London, 1576).

Bullein, William, *The Government of Health* (London, 1595).

Butler, Judith, *Gender Trouble: Gender and the Subversion of Identity* (New York: Routledge, 1990).

Estok, Simon, 'Theory from the Fringes: Animals, Ecocriticism, Shakespeare', *Mosaic*, 40 (2007), 61–78.

Fettiplace, Elinor, *Elinor Fettiplace's Receipt Book*, ed. Hilary Spurling (London: Faber and Faber, 1986).

Fisher, M.F.K., *How to Cook a Wolf* (New York: Farrar, Straus and Giroux, 1942).

Fitzpatrick, Joan, *Food in Shakespeare: Early Modern Dietaries and the Plays* (Burlington, VT: Ashgate, 2007).

Flandrin, Jean-Louis, 'Dietary Choices and Culinary Technique, 1500–1800', in Jean-Louis Flandrin and Massimo Montanari (eds), *Food: A Culinary History from Antiquity to the Present*, tr. Albert Sonnenfeld (New York: Columbia University Press, 1999), 403–417.

———, 'Introduction: The Humanization of Eating Behaviors', in Jean-Louis Flandrin and Massimo Montanari (eds), *Food: A Culinary History from Antiquity to the Present*, tr. Albert Sonnenfeld (New York: Columbia University Press, 1999), 13–20.

Flandrin, Jean-Louis and Massimo Montanari (eds), *Food: A Culinary History from Antiquity to the Present*, tr. Albert Sonnenfeld (New York: Columbia University Press, 1999).

Fudge, Erica (ed.), *Renaissance Beasts: Of Animals, Humans and Other Wonderful Creatures* (Urbana, IL: University of Illinois Press, 2004).

———, 'Saying Nothing Concerning the Same: On Dominion, Purity, and Meat in Early Modern England,' in Erica Fudge (ed.), *Renaissance Beasts* (Urbana: University of Illinois Press, 2004), 70–86.

Fumerton, Patricia and Simon Hunt, *Renaissance Culture and the Everyday* (Philadelphia: University of Pennsylvania Press, 1999).

Hyman, Philip and Mary Hyman, 'Printing the Kitchen, French Cookbooks 1480–1800', in Jean-Louis Flandrin and Massimo Montanari (eds), *Food: A Culinary History from Antiquity to the Present*, tr. Albert Sonnenfeld (New York: Columbia University Press, 1999), 394–402.

Latour, Bruno, *Reassembling the Social: An Introduction to Actor Network Theory* (Oxford: Oxford University Press, 2005).

Mazzio, Carla and David Hillman (eds), *The Body In Parts: Fantasies of Corporeality in Early Modern Europe* (New York: Routledge, 1997).

Schoenfeldt, Michael, 'Fables of the Belly in Early Modern England', in Carla Mazzio and David Hillman (eds), *The Body In Parts: Fantasies of Corporeality in Early Modern Europe* (New York: Routledge, 1997), 243–61.

Scully, Terence, *The Vivendier: A Critical Edition with English Translation* (Totnes: Prospect Books, 1997).

Strong, Roy, *Feast: A History of Grand Eating* (New York: Harcourt, 2002).

Vialles, Noëllie, *Animal to Edible*, tr. J. A. Underwood (Cambridge: Cambridge University Press, 1994).

Wall, Wendy, *Staging Domesticity: Household Work and English Identity in Early Modern Drama* (Cambridge: Cambridge University Press, 2002).

Wecker, John, *Secrets of Nature*, tr. R. Read (London, 1660).

6 Blurred lines

Bestiality and the human ape in Enlightenment Scotland

Andrew Wells[1]

Stories about sexual intercourse between humans and animals are as old as human civilization. Whether mentioned in legal codes (such as the Hittite or Hebrew laws) or as part of ancient mythology (as in Pasiphae's copulation with Poseidon's bull or Zeus's transformation into a swan to seduce Leda), bestiality has, it seems, always been an important element of relations between humans and animals.[2] As such, the relative lack of scholarly attention it has received is surprising, particularly given the sustained interest in matters sexual and bodily that has occupied the academy for the past decades. One recent, otherwise masterly historical survey of sex and the body since 1500, for example, does not even contain entries in its index for bestiality, zoophilia, or animals (either in general or for individual species), and this omission is by no means unique.[3]

Although bestiality was hardly the most common of crimes, such inattention is nonetheless puzzling given the extreme horror and anger that it provoked, as well as its remarkable potential to illuminate aspects of past societies. Alongside pornography and other phenomena (such as monstrous births or disability) usually exiled to the cultural fringes of past and present societies, the study of bestiality provides an opportunity to elucidate both the margins of a given culture (especially the often painfully thin line between what was deemed acceptable and what was not) and its centre. A few scholars have grasped the nettle and dealt with the topic: criminologists, psychologists, and veterinary professionals have concentrated on bestiality as a paraphilia (zoophilia) and on its ethical and psychological dimensions. Historians have usually focused squarely on social or cultural history, especially that of sexuality but also of morality, monstrosity, and witchcraft.[4]

This chapter broadens these efforts by exploring the place of bestiality in both the intellectual history of the (Scottish) Enlightenment and the social history of Enlightenment Scotland.[5] It directs attention towards an oft-noted but little studied thought experiment, which imagined interspecies sexual intercourse between humans and apes as a means of establishing the boundary between humanity and the animal kingdom. The actual performance of this experiment seems to have been inconceivable to all except the German naturalist Eberhard August Wilhelm von Zimmermann (1743–1815), who wrote in 1778 that

> it is ... not entirely improbable that an Ourang [i.e. chimpanzee] could produce an intermediate being with a human. I have heard that such an attempt recently took place in London. A male Ourang was offered a public woman paid for the purpose; however, as far as I know, the attempt turned out fruitless. This experiment was, however, not merely morally objectionable, but physically too. For an equally poor result would probably have happened with such a woman and a human. More-over, it is well known how excessively fierce apes are in copulating, so it presumably spent too soon; such is exactly how this case turned out. In order to undertake something conclusive on this, the attempt must be performed with a man and a female Ourang who would have known one another for some time; that is, if it be allowed and can take place. But this is for moralists to determine.[6]

Three key issues highlighted by this remarkable passage frame the following analysis. First, the gender of the human participant in this experiment was vitally important. The overwhelming majority of acts of bestiality were per-petrated by men, yet a female prostitute was supposedly hired for this exper-iment. Her status as a 'public woman' strongly reinforces the illegitimacy of the attempt, and the proverbial sterility of prostitutes was the implicit reason she would not have conceived by a man, let alone an ape. The first section below examines the reasons women featured in stories of interspecies sex out of all proportion to their actual involvement in acts of bestiality.

The sheer existence of such a remarkable anecdote shows that the back-ground to this experiment also requires close examination. The second section below explores the basis of the thought experiment, especially the connection between the production of fertile offspring and common species identity, and the intellectual commitments of the two enlighten-ment authors – Jean-Jacques Rousseau (1712–78) and James Burnett, Lord Monboddo (1714–99) – who wrote most extensively about it. Finally, the animal participant in this experiment clearly plays a decisive role. The ex-cessive sexual excitement towards a female of 'superior' species led the male ape to climax too soon, and the familiarity of a female ape and male human seemed to bode better for any second attempt. This chapter will proceed to examine actual cases of bestiality prosecuted in the Scottish courts in the seventeenth and eighteenth centuries, paying attention to the importance of animal behaviour and recognising the agency of animals in cases of in-terspecies sex.

Such animal autonomy extended beyond actual intercourse into the thought experiment itself. The notion of 'repugnance' (the resistance dis-played by an animal towards pairing with another) became a likely indicator of species difference in direct response to the difficulties of the interbreeding experiment. The many obstacles in the way of this imagined intercourse thus came to highlight the importance of agency of both animals and humans, not least in making aesthetic and emotional judgements, for establishing species boundaries. These are two of the issues foregrounded in the actual cases of Scottish bestiality to be examined, in which the judgements, emotions, impulses, and acts of animals and humans were of crucial importance in establishing innocence, guilt, remorse, or disgust.

Beyond such continuities between the intellectual and social history of bestiality, comparing its real and imagined variants is valuable for two further reasons. First, it contributes to demonstrating the ongoing importance of Christianity in the midst of potentially sceptical ideas. The notion that humans and apes might be related, and that the extent of this relationship could be established via sexual means, was certainly unorthodox and potentially violated the strict prohibition of human-animal sex (if it emerged that humans and apes were of different species) if not the condemnation of extra-marital intercourse (if it emerged that they were not) in Christian sexual teachings. Second, and related to this, was the importance of such considerations for the history of racial thought. Comparisons between apes and Africans had been made long before this thought experiment was dreamt up, but it will be argued here that heterodox racial ideas were reinforced, not by *imagining* human-ape sex but by *rejecting* it. The decision not to undertake such a lengthy and ethically troubling experiment, while undoubtedly justified on moral grounds, actually foreclosed the one unequivocal and irrefutable proof that all humans are of the same species. A space was thereby left open for racist thinkers and writers to lump together Africans and apes in a shared sexuality and genealogy.

Imagining human-simian sex

Thinking about interspecies sexual intercourse between humans and primates carried with it gendered as well as racial implications. Far the most common configuration of imagined human-ape sex involved a human female and a male ape. Scattered references to such intercourse date from antiquity, but the notion of simian sexual predation on women became increasingly visible over the sixteenth century. Girolamo Cardano (1501–76) and Conrad Geßner (1516–65) popularised accounts of human-ape sex that became prominent in zoological and travel literature, even into the nineteenth century. Edward Topsell's *History of Four-Footed Beasts* (1607), for example, popularised Geßner's *Historia Animalium* (1551–8) in English and repeatedly asserted that apes kidnapped women. The sexual frenzy into which women could drive male apes was even discussed by Darwin in his *Descent of Man* (1871) – albeit discreetly tucked away in a Latin footnote – and the trope probably reached its cultural apogee in the late nineteenth and early twentieth centuries in sculpture, propaganda, and film, especially with the release of *King Kong* (1933; see Figure 6.1).[7]

Figure 6.1 A poster for Merian C. Cooper's 1933 adventure film *King Kong* starring
Fay Wray and Bruce Cabot. Photo by Movie Poster Image Art / Getty
Images.

But the overwhelming majority of actual bestiality cases involved men, so
why were women highlighted in discussions of human–simian sex? Beyond a
generalised conviction that women were closer to nature than men, four reasons for this gender configuration stand out.[8] First, monkeys and apes were
particularly associated with male sexuality. By the Renaissance, this cultural
convention was already old and rested directly on the belief that larger simians would attempt to rape human females when they had the chance. Stories about the unquenchable and furious jealousy of apes when spurned or
betrayed by their human female lovers circulated during the Middle Ages,
inspired in part by Near Eastern tales such as the *Thousand and One Nights*.
In art, apes were commonly used to symbolise lust, often by portraying them
holding mirrors or apples, that fruit of the Fall. The visual tradition associating simians with often illicit or excessive male sexual desire survived into the
eighteenth century, as shown in Figure 6.2. This engraving from 1767 shows
a young girl being 'sacrificed' to the desire of an old and lame suitor; lest
there be any ambiguity about the nature of the scene, in the left background
a monkey is shown molesting a cat.[9]

Second, human female lust often featured prominently in stories about
interspecies sex. Animals could, and did, symbolise the purported inability of women to control their emotions and sexual desires. Indeed, tales of
female bestiality were the obvious extension to loud and chauvinistic attacks on women's supposedly overweening attachment to pets. The fantasy

Figure 6.2 John Collett, *The Sacrifice* (1767), © The Trustees of the British Museum.

of uncoerced sex between women and animals both reinforced female depravity and owed its popularity to placing the male consumer *in loco animalis*. The sexual and emotional were occasionally intertwined, as with the women mourning their shot monkey lovers in *Candide* (1759).[10]

But the fact that many stories explicitly dwelt on sexual activity between women and animals beyond the realm of symbolism is perhaps surprising. After all, not merely did almost all bestiality cases involve (often young) men, but the dictates of early modern jurisprudence meant that bestiality, in common with many other sexual offences, could only be proven if there was clear evidence of penetration.[11] However, the law remained open to the possibility of female bestiality: the English statutes of Henry VIII (1533) and Elizabeth I (1563) referred to 'buggery comyttid with mankynde or beast' without specifying the act or the sex of its perpetrator; the Scottish law of bestiality was based directly on the prohibitions in Leviticus 18 and 20, which dealt equally with men and women; and the law of the Holy Roman Empire, first codified in the *Constitutio Criminalis Carolina* (1533), referred generically to the sexual congress of 'a person with a brute' (*so eyn mensch mit eynem vihe*), a formulation that survived into the late eighteenth century.[12]

The openness of the law to the possibility of female bestiality meant that sex between women and animals, despite being far more common in imagination than in fact, was treated seriously.[13] Of the twelve cases of bestiality tried at the Old Bailey between the 1670s and 1830s, the two with female defendants (1677, 1704) were the most infamous and widely circulated in print. Only one resulted in conviction, but they both involved sex with dogs – the most common type of female-animal sex[14] – and invoked long-standing fears about the 'unnatural wicknedness' of female lust and the possibility of hybrid, monstrous offspring. These attitudes proved to be remarkably persistent. In a pamphlet of 1772, the planter, judge, and proslavery advocate, Edward Long (1734–1813), wrote that 'the lower class of women in *England* ... would connect themselves with horses and asses, if the laws permitted them'.[15] He repeated this two years later, on both occasions to equate miscegenation with bestiality. He even went so far as to repeat Edward Coke's claim that the Henrician statute of 1533 was both introduced and framed to leave open the possibility of female bestiality because a woman had actually conceived by a baboon.[16]

Third, the general demand that female sexuality be passive was combined with an aesthetic hierarchy of sexual attraction to render the non-human participant in supposed cross-species sex almost uniformly male. Thomas Jefferson provided the most concise statement of this idea when he claimed, in his *Notes on the State of Virginia* (1785), that black people recognised their own aesthetic inferiority, giving a 'judgement in favour of the whites, declared by their preference of them, as uniformly as is the preference of the Oranootan for the black women over those of his own species'.[17] Jefferson's analogy with the supposed attraction of the 'Oranootan' for black women made this black-white sexual dynamic perfectly clear: active, male, sexual agents demonstrated the principle that attraction operated vertically, as (male) beings of lesser beauty were attracted to those (females) of greater. Hence the innumerable anecdotes of black women abducted by apes, a trope that survived even into respectable scientific literature of the nineteenth century (see Figure 6.3). (Far fewer are tales of female apes being fond of men, but abduction never takes place in such accounts.)[18]

Jefferson's comments are revealing, not merely because they carry an unmistakable whiff of the white South's frenzied paranoia concerning sexual relations between black men and white women, nor even because his analogical linkage of interspecies and interracial sex restates Edward Long's equation of miscegenation and bestiality. Rather, it is through this analogy that Jefferson alludes to the fourth reason women were prominent in discussions of human-animal sex: their capacity to bear young. The possibility that women abducted by apes had produced offspring was repeatedly discussed, although only the most outlandish of racists persisted in the belief that the origins of black people stemmed from such intercourse.[19] Fictional

Figure 6.3 J. Chapman, 'The Orang–Outang carrying off a Negro Girl' (1795). Frontispiece to Carl Linnaeus, *A Genuine and Universal System of Natural History*, ed. Ebenezer Silby (London, 1795), vol. 2, Wellcome Library, London.

offspring of interspecies liaisons – such as Diderot's 'goat-men' from the *Rêve d'Alembert* (1769) or 'César de Malaca', the fictional child of a human female and a baboon who wrote a lengthy letter 'to others of his species' in Rétif de La Bretonne's *La découverte australe* (1781) – have been versatile tools in the hands of the satirist. But serious scientific study into the possibly of a human-ape hybrid persisted into the twentieth century and cannot yet be ruled out by contemporary science. How much more real must this possibility have seemed in an era in which monstrous births could still be explained as the product of sex with animals?[20]

Belief in this possibility was patchy throughout the early modern period and certainly in decline during the eighteenth century, but not everywhere: it remained alive in popular sex manuals, guide books, satire, and pornography, where even men could be held responsible for producing monstrous hybrid offspring. *Aristotle's Problems* (a question-and-answer work of the early eighteenth century that had nothing to do with Aristotle) recounted a tale in which the birth of a monstrous calf was initially blamed on interspecies sex with a shepherd, before an ill constellation footed the blame. The Marquis de Sade discussed bestiality throughout his writings, and in *The 120 Days of Sodom* (1785) he made it fruitful, but only as a source of further sexual amusement: 'He fucks a goat from behind while being flogged; the goat conceives and gives birth to a monster. Monster though it be, he embuggers it'.[21]

Women remained both the most likely and the most dangerous producers of hybrid offspring. Several 'monsters' produced by female intercourse with animals (including dogs) were described in another pseudo-Aristotelian work, *Aristotle's Masterpiece*, a popular sex guide that remained in print from the 1680s to the 1930s. Ordinarily, the sex of the human perpetrator of bestiality was immaterial from a moral standpoint – the sin was equally as terrible whether committed by a man or a woman – but the possibility of hybrid offspring born to a human made female bestiality much worse. The prospect of a woman carrying the child of an animal posed by far the greater risk to the dignity of humanity, which was predicated upon its divinely ordained mastery of (and hence separation from) the animal realm.[22]

The unthinkable thought experiment

It was precisely to establish the boundary between humanity and the animal world that several eighteenth-century thinkers imagined the experiment in which an ape and human would mate. The enlightenment imperative to understand human nature made delineating the borders of the human species a matter of particular importance. The only concrete method of determining species available in the period was the so-called 'interbreeding criterion', which dictated that two animals are of the same species if they can produce offspring with one another *and* that offspring is fertile. This idea is traceable to Aristotle but was restated in the late seventeenth century by the English botanist and clergyman John Ray (1627–1705) before being immortalised by the Comte de Buffon (1707–88) in the *Histoire naturelle* (1749). In his influential *The Wisdom of God Manifested in the Works of Creation* (1691), Ray 'reckon[ed] all Dogs to be of one *Species* they mingling together in generation, and the breed of such Mixtures being prolifick'.[23]

The uniformity of this principle was of inestimable value to naturalists but remained a puzzle into the nineteenth century:

> why such different Species should not only mingle together, but also generate an Animal, and yet that that hybridous Production should not again generate, and so a new Race be carried on; but Nature should stop here and proceed no further, is ... a Mystery and unaccountable.[24]

Yet this finitude of mixture was ultimately reassuring, since the inability of 'monstrous' hybrid offspring to reproduce and perpetuate their kind prevented the production of 'a new set of Heterogeneous Animals upon Earth'.[25] Exceptions, such as the odd case of a fertile mule, were briskly dealt with: 'I should rather attribute it to a degree of monstrosity in the organs of the mule which conceived, as not being a mixture of two different species, but merely those of either the mare or female ass', wrote John Hunter in 1792.[26]

So why apes? It may now a days seem obvious that the border between humans and the rest of the animal creation is to be sought amongst the higher primates, but this was not immediately apparent before the eighteenth century, especially for those who looked beyond physical characteristics for humanity's distinguishing features. If nobility, bravery, and sporting prowess were the crucial markers of the human, then the horse might seem to be our nearest neighbour (which is one of the reasons discussion of the horse immediately followed that of man in Buffon's *Histoire naturelle*); if reason, industry, and sociability, then beavers, ants, or bees might fit this role. But the possession of reason and a similar physical form induced a range of different classifications of nature (including Ray's and Linnaeus's) to situate apes – or 'anthropomorpha' – next to humanity.[27]

Indeed, these two factors aligned to make it extremely likely that certain apes might well be human. After all, reason was evident through the possession of language, and the ability to speak had a physical manifestation in the form of the larynx. In 1699, the physician and anatomist Edward Tyson (1651–1708) published an account of his dissection of a chimpanzee (which he called a 'pygmie' or 'orang outang'), in which he stated that he 'found the whole Structure' of the larynx 'exactly as 'tis in man. ... And if there was any further advantage for the forming of *Speech*, I can't but think our *Pygmie* had it'.[28] Following the Aristotelian precept that 'nature ... makes nothing in vain', which still guided the interpretation of the natural world, the identical structure of the human and chimpanzee larynx was a powerful indicator that the 'orang outang' possessed reason and thus qualified as human.[29] The belief that certain apes could speak, even if they chose not to, remained state of the art until the Dutch anatomist Pieter Camper (1722–89) debunked the possibility in the pages of the *Philosophical Transactions of the Royal Society* in 1779. For eighty years, therefore, it was possible – even likely – that some form of anthropoid ape might fulfil all the criteria of humanity.[30]

Nevertheless, apes persistently refused to speak, even when tempted with the prospect of immediate baptism. Various explanations were proffered, perhaps the most common of which claimed that apes realised that they would render themselves liable to enslavement if they demonstrated their reason, so they remained silent. The only conclusive means left to settle the question of ape identity was therefore the interbreeding experiment, independently discussed at length by Rousseau and Monboddo. Both of these authors had a vested interest in demonstrating that apes might be human: Rousseau to substantiate his claims about natural man and his idyllic existence, and Monboddo to show that language was neither natural nor God-given but was rather acquired over many years of trial and error. For Rousseau, human apes would show that there was nothing inevitable about the historical development of humankind and that a viable – if now unrealizable – alternative existed; Monboddo, on the other hand, would have been able to point to a human community that had not yet developed language.[31]

Both authors recognised the ethical and practical obstacles that lay in their path. As Rousseau pointed out in his *Discourse on the Origin of Inequality* (1755), in addition to the fact that

> a single generation would not suffice for this experiment, it must be considered impracticable, because it would be necessary for what is only a supposition to be proven true before the experiment that was to prove it true could be tried innocently.[32]

In other words, the interbreeding experiment could take place only once it was certain that humans and apes are of the same species, the very fact that the experiment was intended to prove. (Even this stretches the meaning of 'innocent' to beyond breaking point, since it would involve at least two separate acts of – presumably non-marital – procreation to be conclusive.)

Time, as well as questions of moral causality, stood in the way of the experiment. As Monboddo wrote in 1774,

> it cannot be determined, that any two animals are of the same species till the third generation; so that our knowledge, in this respect, must depend upon a fact which, in all cases, requires a considerable time; and, in many cases, may be very difficult, if not impossible to be ascertained. ... For, though we know, certainly, that [the Ourang Outang] copulates with our females, and though there be the greatest reason to believe, that there is offspring of such copulation, we have no facts by which we can be assured that this offspring will not, like the mule, be barren and unfruitful.[33]

Faced with these insuperable ethical and durational difficulties, several naturalists – such as the German comparative anatomist Johann Friedrich Blumenbach (1752–1840) – largely abandoned the interbreeding criterion to seek alternative markers of species identity in behaviour or morphology. But others simply reconfigured the criterion to focus not on the *consequences* of interspecies sex but on its *causes*. According to this reformulation, the desire (or otherwise) of animals to mate was taken as proof presumptive of species identity or difference; aversion or 'repugnance' to mating indicated species difference, enthusiasm identity.

There were obvious problems with this method, which assumes an unde-viating connection between mutual attraction and the production of both viable and fertile offspring. It was deemed at best unreliable when applied to domesticated species, whose instinctive 'repugnance' had been bred out. Difficulties worsened considerably in the case of human-ape sex, where the avid desire of apes to copulate with women led to a conclusion that few be-sides Rousseau and Monboddo wished to contemplate. Nevertheless, it was the only way to provide near-instant results for the sole accepted, straightfor-ward, and 'natural' test of species.

This is not to say that comprehending animal behaviour was easy. It helped to be working with a familiar species, such as a dog. According to Hunter, this made the best test subject for interbreeding experiments

> as we have opportunities of being acquainted with its disposition and mode of expressing its sensations, which are most distinguishable in the motion of the ears and tail; such as pricking up the ears when anxious, wishing, or in expectation; depressing them when supplicant, or in fear; raising the tail in anger or love, depressing it in fear, and moving it lat-erally in friendship; and likewise raising the hair on the back from many affections of the mind.[34]

He went on to describe the effort to breed a dog with a female wolf, 'which was very tame, and had all the actions of a dog under confinement'. She would not allow any dog to approach her, so she was held and lined twice by a greyhound: 'while in conjunction she remained pretty quiet; but when at liberty, endeavoured to fly at the dog'.[35] While all this points towards resistance that had to be overcome by physical force, the crucial datum is that she conceived, giving birth to four pups. Successful conception, especially in humans, had long been predicated on reciprocal desire, to the extent that cases of rape that resulted in pregnancy were unlikely to be prosecuted successfully. As one writer on medical jurisprudence stated in the 1760s, 'if an absolute rape were to be perpetrated, it is not likely [the victim] would become pregnant'.[36] This belief was on the wane by the second half of the eighteenth century, but it was still strong enough to problematise any interpretation of the female wolf's behaviour: her seemingly unambiguous resistance to being lined by the greyhound was potentially belied by her pregnancy.

The behaviour even of the animal with which man was most familiar, therefore, could prove equivocal and be only imperfectly understood.

Perhaps in response to the drawbacks of looking at potentially unreliable or ambiguous animal desire, efforts were made to restate the closeness of the connection between 'repugnance' and the fully fledged interbreeding criterion. William Lawrence (1783–1867) made the interbreeding criterion itself a response to observed 'repugnance', when he wrote in 1817 that 'the transmission of specific forms by generation, and the aversion to unions with those of other kinds, soon led naturalists to seek for a criterion of species in breeding'.[37] Harriet Ritvo has also described how the lack of issue from a mismatched animal pairing served to *confirm* their initial aversion to one another.[38]

Above all, the use of 'repugnance' as a more-or-less reliable indicator of species identity meant that the problems with the interbreeding criterion were ultimately left for animals themselves to solve. Their desire for or aversion to one other, and enthusiasm or resistance in the face of their (often forced) pairing sometimes served as the only indicators of their species identity. Animals therefore had a substantial degree of agency in determining their own taxonomic fate; the following section will show that they could also have the power of life and death over humans accused of sexual relations with them.

Bestiality in Enlightenment Scotland

Animals had such power simply because the penalties for bestiality were so high in the early modern period. This had not always been the case: the earliest medieval penitentials, for example, treated bestiality on a par with masturbation. But by the high Middle Ages it was viewed as a grave sin, for Aquinas the worst of all kinds of 'unnatural vice'.[39] Despite the efforts of Jeremy Bentham and a few isolated apologists, this attitude survived into the nineteenth century (arguably the present), and by the early modern period, bestiality had become 'the crime not fit to be named among Christians' (*peccatum illud horribile, inter christianos non nominandum*).[40]

The severity of this crime had two important legal consequences. The first was serious disagreement about the punishment for attempted bestiality. In Scottish cases, various authorities were used to argue that the crime was of such a nature that even the attempt (*conatus*) should be punishable by death, at least where the intervention of a third party was the sole reason an attempt had not been realised. Others claimed that the *conatus* was punishable, but surely not to the same extent as when combined with the commission of the act. The Lords of Justiciary ruled in 1719 and again in 1734 that attempted bestiality ought to carry an arbitrary punishment short of death, but in neither of these cases was the attempt relevant: the first defendant (or 'pannel') was found guilty of the act, and the second had a verdict of not proven.[41]

The second consequence was that the ordinary rules of evidence were attenuated, as 'witnesses who are lyable to exceptions will be received, because of the attrocity of the crime'.[42] For the jurist Sir William Blackstone, this augmentation of witnesses was justified because proof 'ought to be the more clear in proportion as the crime is the more detestable'.[43] But he warned that widening the circle of witnesses should not compromise impartiality: bestiality, he wrote,

> is an offence of so dark a nature, so easily charged, and the negative so difficult to be proved, that the accusation should be clearly made out: for, if false, it deserves a punishment inferior only to that of the crime itself.[44]

Punishment for false accusation did not materialise in the case of John Rainnie, who was accused of committing bestiality with a ewe in 1718. Although the suit against him was abandoned, this must have been cold comfort, as he had explicitly requested that the matter proceed to trial, given that 'ane accusation of this nature does generally tho the proof should faile leave a staine in the Character of any person who is so unhappy as to be Charged with it'.[45]

Living under this suspicion was perhaps preferable to facing the vagaries of prosecution evidence in a trial. In more than one instance, animal behaviour was explicitly adduced as proof of interspecies intercourse. In the case of seventeen-year-old Thomas Wisheart, who was tried in 1691 for bestiality with a mare, both the key witness and the indictment ('libel') stipulated that 'after ye parted with the mear she was sein [seen to] cast up her tail and cast out as mears uses to doe when ston'd horss ['stoned horses', i.e. stallions] part with them'. One witness against George Robertson in 1719 stated that the 'posture and Condition' of the mare with which he was accused of copulating was 'as a mare is in when she is covered with a stone[d] horse'. Another testified that the mare's tail stood out stiff from its body. In these cases, which both resulted in conviction and execution, no eyewitness testified to observing penetration take place; rather, their testimony focussed on the motions of the accused during copulation and his dishevelled appearance upon disengaging from the animal. The immediate postcoital behaviour of both participants, human and animal, was also recorded. The fact that no concrete evidence of penetration was offered but that animal comportment was mentioned in both indictments and witness testimony suggests that it had considerable evidentiary weight.[46]

Animals also exercised agency in more direct ways, most often by resisting their sexual assault at human hands. When John White was tried in 1699 for bestiality with a bitch, one witness testified that she saw 'him copulat with the sd bitch, and that she did see him three tymes reingadge the bitch after the bitch strugled to be away'. This circumstance was explicitly mentioned in White's indictment. Both witnesses in the case of Robert Howie (1717) testified to the reaction of the cow he was assaulting: they both saw 'the Cow throw the pannel from her in the grub of the byre and the pannel rose and went again to the Cow'. The survival of such details in the legal record – especially in indictments – suggests that the persistence of the offender in the face of resistance was an aggravation of their guilt.[47]

But animals could also behave in a manner that was interpreted as affectionate or even defensive towards their human assailant. John Burton, a witness in Robertson's trial, testified that as he approached Robertson in flagrante delicto, 'he saw the mare turn and fann [fawn] upon the pannel and that when he the Deponent came up and offered to take holed of the mare she offered to kick at him and run away'. Of course, such behaviour was used to prove that bestiality had taken place, perhaps nowhere as spectacularly as in the 1677 Old Bailey trial, in which a woman was convicted of 'Buggery with a certain Mungril Dog'. The dog was actually brought to court, and 'being set on the Bar before the Prisoner, owned her by wagging his tail, and making motions as it were to kiss her, which 'twas sworn she did do when she made that horrid use of him'. The attendance at court of such an animal 'witness' shows the lengths to which clear proof of bestiality was sought: the entire spectrum of animal behaviour – from resistance to affection – was interpreted to support its prosecution.[48]

The influence of (sex with) animals on humans, particularly on their emotional state, was also not inconsiderable. Several witnesses spoke of their fear or disorientation at having observed acts of interspecies sex: Jean Brodie 'was so astonished that she fell into a swound [swoon] for a whill', and Arthur Craig, 'being exceedingly … astonished he fell down upon ye ground at ye back of a whine buss [gorse bush] upon his knees & said Lord have mercy on me for ye sight I have seen'. Craig was so affected that he did not inform anybody until the following evening, 'his Master expostulating wt him Because of ye dejection & sadness [which] he observed him to be under [that] day'. Perpetrators were often described as 'ashamed', 'confused', or 'confounded' by their acts, and they frequently ignored witnesses who addressed or challenged them at the scene of the crime.[49]

The clearest example of emotional distress is provided by the case of Robert Beggs. Beggs was a married soldier in the 6th Regiment of Foot, based in Fraserburgh, who was caught by his fellow soldier John Cuming having sex with a cow on a moonlit night in August 1749. Cuming immediately left to report the incident to a superior; he was waiting in the guardroom when Beggs returned. Confronted by Cuming, Beggs collapsed in tears. Over the course of the evening, he twice fell to his knees and begged first Cuming and then their sergeant for forgiveness. On both occasions he mentioned the shame of the crime, which was 'so Monstruous, that he would rather dye than be Wend to it, or Accused of it', and remarkably he even claimed 'that it might be a Mean[s] of Making him a better Christian than hitherto he had been'. This was not entirely persuasive. Cuming later brought the cow to the guardroom, 'and the pannell Begg'd for God's Sake that the Cow might be taken away', which was done. The case eventually came to trial in January 1750 and was found not proven.[50]

Bestiality clearly provoked strong emotions in all involved, witnesses, perpetrators, and – judging from the efforts of some to resist – victims.[51] The shame evoked by the deed extended to God-fearing spectators, and the later presence of the animal could produce a powerful response, especially in its assailant. The sense of shame and fear produced by interspecies sex can be glimpsed in the legal record, which frequently described the sexual acts of bestiality in animal rather than human terms. Perhaps unsurprisingly for a crime that was 'not to be named', descriptions of such sex were often deliberately euphemistic: witnesses described perpetrators as 'working too and froe ... as a st[o]ned horse uses to do to a mear', or acting 'as a horse uses to Cover a mare', and one testified that 'the motions he saw were like the motions of nature and such as are used in Carnall Dealing'. Such descriptions situated this behaviour firmly in the animal, natural realm and removed the perpetrator from the human community. Indeed, it was so 'that no vestige of them might be committed to our common parent' that those convicted of bestiality in Scotland were sentenced to be strangled then burnt at the stake. And the memory of this linkage of shame, fear, and animality was to be obliterated by destroying the animal, something also commanded by Leviticus 20:15.[52]

Conclusion: bestiality and race – a missed opportunity?

Such an emotional and legal context must have been a powerful disincentive to undertaking the interbreeding experiment, which was ultimately never attempted. Zimmermann's 1778 anecdote – which he later claimed to have heard from an English officer – raised many eyebrows and prompted the German naturalist Peter Simon Pallas (1741–1811) to write to his English correspondent Thomas Pennant, asking 'is this really fact & have not any other such trials been made on a better plan in England or on the African coast (where it would be more easy) & with what success?' Pennant's answer was published by Pallas in his *Neue Nordische Beyträge*: 'Professor *Zimmermann* was imposed on by the account of the commerce between the Orang outăng and the prostitute in London. We are a wicked people, but not so bad, as to permit so infamous a deed.' Had the event described by Zimmermann actually taken place, Pallas observed, there were several reasons it would not have been successful, amongst which was the climate, described as 'disadvantageous and deadly' for Orang-Outangs.[53] He nevertheless thought Pennant would agree with him that it was desirable to have some genuine observations concerning the mixed offspring of a human and ape,

> or that corrupt Europeans in the [African or West] Indian plantations would undertake something which, in contrast to other vices, would at least be instructive for natural history, and in a hot climate arrange for a bastard offspring of an Orang-utang and a female slave for European observers.[54]

Pallas's comments tear the veil from the interbreeding experiment to expose its racist foundations. These were long-standing, as his repeated emphasis on climate makes clear. This had long been cited as the basis of physical and moral diversity amongst the world's peoples, and variations in both skin colour and the handling of bestiality were attributed to climatic differences. As the Scottish jurist Sir George Mackenzie stated in 1678, 'in hotter Countreys, where Custome, and Climat, lessens this Crime, the Crime is by their Lawyers thought punishable less severely'.[55] The association of crime and clime was well established and emerged in such disparate genres as sermons and erotica. Few, however, matched the vitriol of Edward Long, whose discussion of bestiality openly bridged the gap between climate and race:

> An example of this intercourse once happened, I think, in England; and if lust can prompt to such excesses in that Northern region, and in despight of all the checks which national politeness and refined sentiments impose, how freely may it not operate in the more genial soil of Afric, that parent of every thing that is monstrous in nature.[56]

Long's arguments were clearly and offensively racist. But the studious avoidance of human diversity in other discussions of the interbreeding experiment belied the racism they also contained. Monboddo and Rousseau could potentially be viewed as anti-racist for seeking to broaden the definition of humanity to such an extent that the orang-outang was contained within it. But Long sought to do the same (albeit separating 'white men' from 'the Negroe race' and the 'oran-outang' by creating two species within one 'human *genus*'), and one need look no further than the frequent use of tales relating the simian abduction of *African* women by these authors to recognise that their apparent colour-blindness is nothing of the sort.[57]

Such tales of abduction highlight the role played by animal agency – real or imagined – in both scientific writing and courtroom testimony. The capacity of animals to determine ideas and events is manifest throughout the social history of bestiality and the intellectual history of the interbreeding experiment. An ape's refusal to speak despite the requisite anatomy forced humans to imagine the sexual clarification of species boundaries. Aversion to mixture with other animals might indicate a separate species identity, but this was never an ironclad principle. Resisting sexual assault at the hands of a human could reinforce the case against the accused, but so too could affection, whether displayed in the courtroom or at the scene of the crime. Animals and the cruelty to which they were subjected were also able to evoke a range of emotions in bystanders, perpetrators, and the reading and writing public. Ultimately, humans did their best to interpret animal behaviour in accordance with their own designs, but animals did not always make it easy for them to do so.

This was even the case in the one context where humans undeniably controlled every aspect of human-animal interaction: in the imagination. This was the only location where the interbreeding experiment could take place, yet the actual behaviour of animals made it impossible for thinkers to predict its outcome. But while the whole notion of a human-ape hybrid remains an extremely delicate ethical prospect, the experiment, had it been undertaken, might have produced one positive result. If the attempt were met with no offspring, it would have been absolutely undeniable that apes and humans are different species, and the racist potential of bestiality – even of the entire human-ape analogy – would have been wounded, perhaps mortally. However, given race's hydra-like capacity to survive despite repeated and near-fatal blows, this counterfactual prognosis is almost certainly too optimistic.

Notes

1 The author would like to thank Katrin Berndt, Sarah Cockram, Erica Fudge, Harriet Ritvo, Silvia Sebastiani, and the anonymous reviewers for their careful reading and insightful comments. Thanks also go to the National Records of Scotland and Sir Robert Clerk of Penicuik for permission to reproduce material and to the Leverhulme Trust and the Graduiertenschule für Geisteswissenschaften Göttingen for financial support during the research and preparation of this essay.
2 Harry A. Hoffner, Jr., 'Hittite Laws', in Martha T. Roth (ed.), *Law Collections from Mesopotamia and Asia Minor* (Atlanta: Scholars Press, 1995), 236–37; Apollodorus, *The Library of Greek Mythology*, tr. Robin Hard (Oxford: Oxford University Press, 1997), III.1.3, III.10.7; Roland Boer, 'From Horse Kissing to Beastly Emissions: Paraphilias in the Ancient Near East', in Mark Masterson, Nancy Sorkin Rabinowitz, and James Robson (eds), *Sex in Antiquity: Exploring Gender and Sexuality in the Ancient World* (London: Routledge, 2015), 67–79.
3 Sarah Toulalan and Kate Fisher (eds), *The Routledge History of Sex and the Body, 1500 to the Present* (London: Routledge, 2013).
4 See also David Cressy, *Agnes Bowker's Cat: Travesties and Transgressions in Tudor and Stuart England* (Oxford: Oxford University Press, 2000), 7–8; Feona Attwood, 'Reading Porn: The Paradigm Shift in Pornography Research', *Sexualities*, 5 (2002), 91–105; Andrea M. Beetz and Anthony L. Podberscek (eds), *Bestiality and Zoophilia: Sexual Relations with Animals*, special issue of *Anthrozöos* (2005); Piers Bierne, 'Rethinking Bestiality: Towards a Concept of Interspecies Sexual Assault', *Theoretical Criminology*, 1 (1997), 317–40; Graham Parker, 'Is a Duck an Animal? An Exploration of Bestiality as a Crime', *Criminal Justice History*, 7 (1986), 95–109; Erica Fudge, 'Monstrous Acts: Bestiality in Early Modern England', *History Today*, 50/8 (2000), 20–25; John M. Murrin, '"Things Fearful to Name": Bestiality in Early America', in Angela N. H. Creager and William Chester Jordan (eds), *The Animal-Human Boundary: Historical Perspectives* (Rochester: University of Rochester Press, 2002), 115–56; Jens Rydström, *Sinners and Citizens: Bestiality and Homosexuality in Sweden, 1880–1950* (Chicago: University of Chicago Press, 2003); Courtney Thomas, '"Not Having God Before his Eyes": Bestiality in Early Modern England', *Sixteenth Century*, 26 (2011), 149–73; Doron S. Ben-Atar and Richard D. Brown, *Taming Lust: Crimes Against Nature in the Early Republic* (Philadelphia: University of Pennsylvania Press, 2014).
5 Existing research on bestiality in Scotland is limited to P. G. Maxwell-Stuart, '"Wild, filthie, execrabill, detestabill, and unnatural sin": Bestiality in Early Modern Scotland', in Tom Betteridge (ed.), *Sodomy in Early Modern Europe* (Manchester: Manchester University Press, 2002), 82–93 and a forthcoming article by Anne-Marie Kilday in *Journal of the History of Sexuality*. For an illuminating survey of the discourse on bestiality in the French Enlightenment, see Vincent Jolivet, 'Lumières et bestialité', *Dix-huitième siècle*, 42 (2010), 285–303. Thanks to Silvia Sebastiani for sharing this reference.

6   'Es ist ... nicht ganz unwahrscheinlich, daß ein Ourang mit einem Menschen eine Mittelgattung hervorbringen könnte. Ich habe erfahren, daß man vor kurzem in London einen solchen Versuch angestellet hat. Man bot dem männlichen Ourang eine dafür bezahlte öffentliche Weibsperson an; allein, so viel ich weiß, lief der Versuch ganz fruchtlos ab. Dieser Versuch war aber nicht nur moralisch verwerflich, sondern auch physikalisch. Denn ein eben so schlechter Erfolg würde sich wahrscheinlich mit einer solchen Weibsperson auch durch Zuthun eines Menschen ereignet haben. Ueberdem weiß man ja, wie übermäßig heftig die Affen bey der Begattung sind, welches von selbst eine zu frühzeitige Verschwendung vermuthen ließ; gerade so traf auch der Fall ein. Um irgend etwas entscheidendes hierunter anzustellen, so mußte dieser Versuch mit einem Manne und Ourangs-Weibchen vorgenommen werden, welche sich seit einiger Zeit gekannt hätten, falls es je erlaubt ist und seyn kann; denn dies lasse ich den Moralisten zu untersuchen über'. E. A. W. Von Zimmermann, *Geographische Geschichte des Menschen*, 3 vols (Leipzig, 1778–83), i. 118n. All translations are the author's own.

7   H. W. Janson, *Apes and Ape Lore in the Middle Ages and the Renaissance* (London: Warburg Institute, 1952), 267–68, 270; Edward Topsell, *The Historie of Foure-Footed Beasts* (London, 1607), 2–15 passim; Charles Darwin, *The Descent of Man* (London: Penguin, 2004), 25 n. 11.

8   On this conviction see Margaret R. Sommerville, *Sex and Subjection: Attitudes to Women in Early Modern Society* (London: Arnold, 1995), 8–16; Merry E. Wiesner, *Women and Gender in Early Modern Europe* (2nd ed., Cambridge: Cambridge University Press, 2000), 33.

9   Janson, *Apes and Ape Lore*, 271–72, 285 n. 88; Irene Earls, *Renaissance Art: A Topical Dictionary* (Westport: Greenwood Press, 1987), 262 (s.v. 'Seven Deadly Sins').

10 Ingrid H. Tague, *Animal Companions: Pets and Social Change in Eighteenth-Century Britain* (University Park: State University of Pennsylvania Press, 2015), chapter 3; Midas Dekkers, *Dearest Pet: On Bestiality*, tr. Paul Vincent (London: Verso, 2000), 154–61; Janson, *Apes and Ape Lore*, 261–67; Voltaire, *Candide and Other Stories*, tr. Roger Pearson (Oxford: Oxford University Press, 1990), 40–41; Jolivet, 'Lumières', 290–91.

11 Only 14 females were tried for bestiality in Sweden between 1635 and 1754 out of a total of 1500 cases, an astronomical amount given that the Old Bailey saw only 12 trials for bestiality and 97 for 'sodomy' between 1677 and 1830. See Jonas Liliequist, 'Peasants against Nature: Crossing the Boundaries between Human and Animal in Seventeenth- and Eighteenth-Century Sweden', *Journal of the History of Sexuality*, 1 (1991), 395; *Old Bailey Proceedings Online* (www.oldbaileyonline.org, version 7.2, 08 January 2016) [hereafter *OBSP*]: t16770711-1, t16770711-2 (both 1677), t17040426-42 (1704), t17570713-29 (1757), t17701024-38 (1770), t17760417-28 (1776), t17790915-50 (1779), t17870100-64 (1787), t17991030-84 (1799), t18200918-65 (1820), t18271206-3 (1827), t18300218-13 (1830).

12 William Blackstone, *Commentaries on the Laws of England*, 4 vols (Oxford, 1765–9), iv. 215; 25 Hen. VIII, c. 6; 5 Eliz. I, c. 17; Lev. 18: 23, 20: 15–16; *Des allerdurchleuchtigsten großmechtigste unüberwindtlichsten Keyser Karls ... peinlich gerichts ordnung* (Mainz, 1533), § 116; *Constitutio Criminalis Theresiana* (Vienna, 1769), art. 74 § 1.

13 On the (male) imagination of female bestiality, see Alfred C. Kinsey, Wardell B. Pomeroy, Clyde E. Martin, and Paul H. Gebhard, *Sexual Behavior in the Human Female* (Philadelphia: W. B. Saunders, 1953), 502; Erica Fudge, *Perceiving Animals: Humans and Beasts in Early Modern English Culture* (Basingstoke: Macmillan, 2000), 201 n. 97.

14 See Kinsey, *Human Female*, 506 (and n. 7); Hani Miletski, *Understanding Besti-ality and Zoophilia* (Bethseda, MD: East-West Publishing, 2002), 131. For Moll Flanders, sex with a dog was almost preferable to fraternal incest. See Daniel Defoe, *Moll Flanders*, ed. G. A. Starr (Oxford: Oxford University Press, 1971), 98.

15 'A Planter' [Edward Long], *Candid Reflections Upon … The Negroe-Cause* (London, 1772), 48–49 (emphasis in original).

16 *OBSP*, July 1677, trial of married woman (t16770711-1) and April 1704, trial of Mary Price, alias Hartington (t17040426-42); *A True Narrative of the Proceedings at the Sessions-House in the Old-Baily: Beginning on the 11th of … July, 1677* (London, 1677), 1; *News from Tybourn: Being An Account of the Confession and Execution of the Woman condemned for committing that horrid sin of Buggery with a Dog* (London, 1677); *A Compleat Collection of Remarkable Tryals of the most Notorious Malefac-tors at the Sessions-House in the Old Baily*, 4 vols (London, 1718–21), ii. 94–95; [Edward Long], *The History of Jamaica*, 3 vols (London, 1774), ii. 383 and n.; Sara Salih, 'Filling Up the Space Between Mankind and Ape: Racism, Speciesism and the Androphilic Ape', *Ariel: A Review of International English Literature*, 38 (2007), 95–111; Edward Coke, *The Third Part of the Institutes of the Laws of England* (London, 1644), 59.

17 Thomas Jefferson, *Notes on the State of Virginia* (2nd American ed., Philadelphia, 1794), 201.

18 Examples include: John Ogilby, *Africa* (London, 1670), 558; [Georges-Louis Leclerc, Comte de Buffon], *Histoire naturelle, générale et particulière*, 15 vols (Paris, 1749–67), xiv. 50–51; James Burnett, Lord Monboddo, 'Of the Ourang Outang & whether he be of the Human Species' and 'Letter from the Bristol Merchant con-cerning the Oran-Outang', National Library of Scotland (hereafter NLS), MS 24537, fos. 18r, 40r; Baron [Georges] Cuvier, *The Animal Kingdom, Arranged after its Organization*, tr. and ed. Edward Blyth et al. (London: William S. Orr and Co., 1854), 56; (on female apes' fondness of men) M. de Blainville, *Travels through Holland, Germany, Switzerland, and Other Parts of Europe*, 2 vols (London, 1743), i. 259.

19 See, for example, John Gardner Kemeys, *Free and Candid Reflections Occasioned by the Late Additional Duties on Sugars and on Rum* (London, 1783), 72–79 n. Kemeys also wrote to Long to inform him that one of the women on his plantation 'bears the strongest Marks of a mixture', she having 'the most perfect resemblance to the Baboon tribe'. Kemeys to Long, 20 July 1784. British Library, Additional MS 12431, fo. 205r.

20 Miriam Claude Meijer, *Race and Aesthetics in the Anthropology of Petrus Camper (1722–1789)* (Amsterdam: Rodopi, 1999), 123–27; Denis Diderot, *Œuvres philosophiques*, ed. Paul Vernière (Paris: Garnier, 1964), 381–84; [Rétif de la Bretonne], *La Découverte australe Par un Homme-volant*, 3 vols (Leipzig, 1781), iii. 1–92; Kirill Rossiianov, 'Beyond Species: Il'ya Ivanov and His Experiments on Cross-Breeding Humans with Anthropoid Apes', *Science in Context*, 15 (2002), 277–316; Russell H. Tuttle, *Apes and Human Evolution* (Cambridge: Harvard University Press, 2014), 33–34; Dekkers, *Dearest Pet*, 88–91.

21 Lorraine Daston and Katharine Park, *Wonders and the Order of Nature* (New York: Zone, 2001), 192; *Aristotle's Book of Problems* (27th ed., London, 1720), 67; Marquis de Sade, 'Philosophy in the Bedroom', in Richard Seaver and Austryn Wainhouse (ed.), *Justine, Philosophy in the Bedroom, and Other Writings* (New York: Grove Press, 1965), 328–29; id., *Juliette*, ed. Austryn Wainhouse (New York: Grove Press, 1968), 188, 744–47; id., *The 120 Days of Sodom and Other Writings*, ed. Richard Seaver and Austryn Wainhouse (New York: Grove Press, 1966), 603; Jolivet, 'Lumières', 298–301; Julia V. Douthwaite, *The Wild Girl, Natural Man, and the Monster: Dangerous Experiments in the Age of Enlightenment* (Chicago: University of Chicago Press, 2002), 202.

22 *Aristotle's Master-Piece: Or, the Secrets of Generation* (London, 1690), 52, 177; *Aristotle's Master-Piece Completed* (Glasgow, 1781), 33; Keith Thomas, *Man and the Natural World* (London: Penguin, 1984), 17–50 (esp. 39).

23 John Ray, *The Wisdom of God Manifested in the Works of the Creation* (London, 1691), 5.

24 Ray, *Wisdom*, 219; James Cowles Prichard, *Researches into the Physical History of Man*, ed. George W. Stocking, Jr. (Chicago: University of Chicago Press, 1973 [1813]), 7–13.

25 James Parsons, *A Mechanical and Critical Enquiry into the Nature of Hermaphrodites* (London, 1741), 7; Harriet Ritvo, 'Barring the Cross: Miscegenation and Purity in Eighteenth- and Nineteenth-Century Britain', in Diana Fuss (ed.), *Human, All Too Human* (New York: Routledge, 1996), 42.

26 John Hunter, *Observations on Certain Parts of the Animal Oeconomy* (2nd ed., London, 1792), 143.

27 Thomas, *Man and the Natural World*, 100–101; [Buffon], *Histoire naturelle*, vols 2 and 3; Oliver Goldsmith, *History of the Earth and Animated Nature*, 8 vols (2nd ed., London, 1779), ii. 328; Jonathan Swift, *Gulliver's Travels*, ed. Claude Rawson (Oxford: Oxford University Press, 2005), 116; [James Burnett, Lord Monboddo,] *Of the Origin and Progress of Language* [hereafter *OPL*], 6 vols (Edinburgh, 1773–96), i. 211–12; Philip Sloan, 'The Gaze of Natural History', in Christopher Fox, Roy Porter, and Robert Wokler (eds), *Inventing Human Science: Eighteenth-Century Domains* (Berkeley: University of California Press, 1995), 112–51.

28 Edward Tyson, *Orang-Outang, sive Homo Sylvestris: Or, the Anatomy of a Pygmie* (London, 1699), 51.

29 Aristotle, *Politics*, 1253a8.

30 Peter Camper, 'Account of the Organs of Speech of the Orang Outang', *Philosophical Transactions of the Royal Society*, 69, 1 (1779), 139–59. See also Silvia Sebastiani, 'Challenging Boundaries: Apes and Savages in Enlightenment', in Wulf D. Hund, Charles W. Mills, and Silvia Sebastiani (eds), *Simianization: Apes, Gender, Class, and Race* (Racism Analysis Yearbook, 6: 2015/16; Zürich: Lit, 2015), 105–37.

31 Diderot, *Œuvres philosophiques*, 385 (on the promise of baptism). The idea that the Orang-Outang stayed silent in order to remain free seems to have originated with a Dutch physician in Java, Jacob de Bondt. See Miriam Claude Meijer, 'Bones, Law, and Order in Amsterdam: Petrus Camper's Morphological Insights', in Klaas van Berkel and Bart Ramakers (eds), *Petrus Camper in Context: Science, The Arts, and Society in the Eighteenth-Century Dutch Republic* (Hilversum: Verloren, 2015), 206; Jacobi Bontii [Jacob de Bondt], *Historiæ Naturalis et Medicæ Indiæ Orientalis*, addendum to Gulielmi Pisonis [Willem Piso], *De Indiæ Utriusque Re Naturali et Medica* (Amsterdam, 1658), 56.

32 'Jean Jaques Rousseau, Citoyen de Genève', *Discours sur l'origine et les fondements de l'inégalité parmi les hommes* (Amsterdam, 1755), 229–30.

33 Monboddo, *OPL* (2nd ed., Edinburgh, 1774), i. 334.

34 Hunter, *Observations*, 148.

35 Hunter, *Observations*, 148.

36 [Johann Friedrich Faselius,] *Elements of Medical Jurisprudence*, tr. Samuel Farr (London, 1788), 43.

37 William Lawrence, *Lectures on Physiology, Zoology, and the Natural History of Man* (London, 1819), 265.

38 Ritvo, 'Barring the Cross', 41.

39 Aquinas, *Summa Theologica*, II-II, q. 154 a. 12.

40 Blackstone, *Commentaries*, iv. 216; Jeremy Bentham, 'Sextus', in Philip Schofield, Catherine Pease-Watkin, and Michael Quinn (ed.), *Of Sexual Irregularities, and Other Writings on Sexual Morality* (Oxford: Oxford University Press, 2014), 98.

41 National Records of Scotland, High Court of Justiciary Records (here-after NRS, HCJ): JC7/9 (Minute Book, 1718–20), pp. 570–77; David Hume, *Commentaries on the Law of Scotland, Respecting Crimes*, 2 vols (2nd ed., Edinburgh, 1819), i. 27, 466.

42 Sir George Mackenzie, *The Laws and Customes of Scotland, in Matters Criminal* (Edinburgh, 1678), 161.

43 Blackstone, *Commentaries*, iv. 215.

44 Blackstone, *Commentaries*, iv. 215.

45 NRS, HCJ: JC7/9, pp. 348, 353.

46 NRS, HCJ: JC6/13 (Minute Book, 1690–95), p. 335; JC26/73 (Processes, 1691), 'Indytment agt Thomas Wisheart for Bestialitie, 1691'; JC7/9, pp. 578–80.

47 NRS, HCJ: JC6/14 (Minute Book, 1695–1701), fo. 229r; JC2/19 (Book of Adjournall, 1693–9), fo. 448v; JC7/8 (Minute Book, 1716–18), p. 406.

48 NRS, HCJ: JC7/9, p. 579; *True Narrative*, 1–2.

49 NRS, HCJ: JC6/13, pp. 335–6; NRS, Clerk of Penicuik Papers: GD18/698/1 (Deposition of Arthur Craig, 22 August 1705); NRS, HCJ: JC7/5 (Minute Book, 1711–12), pp. 542–43; NLS, Erskine Papers, MS 5073 (Correspondence, 1726–31), fo. 35r. Thanks to Sir Robert Clerk of Penicuik for permission to quote from this material.

50 NRS, HCJ: JC7/27 (Minute Book, 1747–51), pp. 310–26.

51 On the interpretation of animal behaviour to discern emotions, see the instructive survey by Marian Dawkins, 'Animal Welfare and the Paradox of Animal Consciousness', *Advances in the Study of Behavior*, 47 (2015), 5–38. See also Charles Darwin, *The Expression of the Emotions in Man and Animals* (London: John Murray, 1872) and Jaak Panksepp, 'The Basic Emotional Circuits of Mammalian Brains: Do Animals Have Affective Lives?', *Neuroscience and Behavioral Reviews*, 35 (2011), 1791–1804.

52 Coke, *Institutes*, 58–59; NRS, HCJ: JC6/13, p. 335; JC7/9, pp. 578–79; JC7/19, pp. 84–85; Hume, *Commentaries*, i. 466.

53 'nachtheiligen und tödtlichen'. *Neue Nordische Beyträge*, 1 (1781), 156.

54 'oder daß auch verdorbne Europäer in den dortigen oder indianischen Pflan-zorten einmal statt andrer Laster dieses, wenigstens für die Thiergeschichte unterrichtende begiengen, und in einem heißen Klima eine Bastarterzeugung vom Orang-utang und einer Sklavinn für europäische Beobachter besorgten'. Ibid., 156. See also Zimmermann, *Geographische Geschichte*, vol. iii, sig. a [4]; Petra Feuerstein-Herz, 'Eberhard August Wilhelm von Zimmermann (1743–1815) und die Tiergeographie', Diss. (Technische Universität Braunschweig, 2003), 292–97; Carol Urness (ed.), *A Naturalist in Russia: Letters from Peter Simon Pallas to Thomas Pennant* (Minneapolis: University of Minnesota Press, 1967), 128. For a wider discussion of German thought on the relation of man and ape in the eighteenth century, see Carl Niekerk, 'Man and Orangutan in Eighteenth-Century Thinking: Retracing the Early History of Dutch and German Anthropology', *Monatshefte für Deutschsprachige Literatur und Kultur*, 96 (2004), 477–502.

55 Mackenzie, *Laws and Customes*, 161.

56 [Long], *Jamaica*, ii. 383; [Richard Smalbroke], *Reformation Necessary to Prevent Our Ruine* (London, 1728), 21; [Thomas Stretser], *A New Description of Merryland* (4th ed., Bath, 1741), 7.

57 In addition to his published utterances on the subject, Long's notes for an unpublished second edition of the *History* discussed 'negro'-ape hybrids at great length. See British Library, Additional MS 12405, esp. fos. 280r–282v, 288r–293v.

Bibliography

Primary sources

Manuscript sources

British Library, London
  Additional Manuscripts
    12405 Annotated copy of Long's *History of Jamaica*, vol. 2.
    12431 Papers relating to Jamaica

National Records of Scotland, Edinburgh
  High Court of Justiciary Records (JC)
    JC2 Books of Adjournall
    JC6, JC7 Minute Books
    JC26 Processes
  Clerk of Penicuik Papers (GD18)
    GD18/698 Papers regarding an accusation of bestiality

National Library of Scotland, Edinburgh
  Monboddo Papers
    MS 24537 Notes and Essays
  Erskine Papers
    MS 5073 Correspondence, 1726–31

Printed sources

Anon., *Aristotle's Book of Problems* (27th ed., London, 1720).

Anon., *Aristotle's Master-Piece Completed* (Glasgow, 1781).

Anon., *Aristotle's Master-Piece: Or, the Secrets of Generation* (London, 1690).

Anon., *A Compleat Collection of Remarkable Tryals of the Most Notorious Malefactors at the Sessions-House in the Old Baily*, 4 vols (London, 1718–21).

Anon., *News from Tybourn: Being an Account of the Confession and Execution of the Woman Condemned for Committing that Horrid Sin of Buggery with a Dog* (London, 1677).

Anon., *A True Narrative of the Proceedings at the Sessions-House in the Old-Baily: Beginning on the 11th of … July, 1677* (London, 1677).

Apollodorus, *The Library of Greek Mythology*, tr. Robin Hard (Oxford: Oxford University Press, 1997).

Aquinas, *St. Thomas Aquinas on Politics and Ethics*, tr. and ed. Paul E. Sigmund (New York: Norton, 1988).

Aristotle, *The Politics and the Constitution of Athens*, ed. Stephen Everson (Cambridge: Cambridge University Press, 1996).

Bentham, Jeremy, *Of Sexual Irregularities, and Other Writings on Sexual Morality*, ed. Philip Schofield, Catherine Pease-Watkin, and Michael Quinn (Oxford: Oxford University Press, 2014).

Blackstone, William, *Commentaries on the Laws of England*, 4 vols (Oxford, 1765–9).

Bontii, Jacobi [Jacob de Bondt], *Historiæ Naturalis et Medicæ Indiæ Orientalis*, addendum to Pisonis, Gulielmi [Willem Piso], *De Indiæ Utriusque Re Naturali et Medica* (Amsterdam, 1658).

[Buffon, Georges-Louis Leclerc, Comte de], *Histoire naturelle, générale et particulière*, 15 vols (Paris, 1749–67).

[Burnett, James, Lord Monboddo,] *Of the Origin and Progress of Language*, 6 vols (Edinburgh, 1773–96).

Camper, Peter, 'Account of the Organs of Speech of the Orang Outang', *Philosophical Transactions of the Royal Society*, 69, 1 (1779), 139–59.

Coke, Edward, *The Third Part of the Institutes of the Laws of England* (London, 1644).

Constitutio Criminalis Theresiana (Vienna, 1769).

Cuvier, Baron [Georges], *The Animal Kingdom, Arranged after Its Organization*, tr. and ed. Edward Blyth et al. (London: William S. Orr and Co., 1854).

Darwin, Charles, *The Expression of the Emotions in Man and Animals* (London: John Murray, 1872).

————, *The Descent of Man* (London: Penguin, 2004).

de Blainville, M., *Travels through Holland, Germany, Switzerland, and Other Parts of Europe*, 2 vols (London, 1743).

Defoe, Daniel, *Moll Flanders*, ed. G. A. Starr (Oxford: Oxford University Press, 1971).

[De la Bretonne, Rétif], *La Découverte australe Par un Homme-volant*, 3 vols (Leipzig, 1781).

Des allerdurchleuchtigsten großmechtigste unüberwindtlichsten Keyser Karls … peinlich gerichts ordnung (Mainz, 1533).

de Sade, Marquis, 'Philosophy in the Bedroom', in *Justine, Philosophy in the Bedroom, and Other Writings*, ed. Richard Seaver and Austryn Wainhouse (New York: Grove Press, 1965).

————, *The 120 Days of Sodom and Other Writings*, ed. Richard Seaver and Austryn Wainhouse (New York: Grove Press, 1966).

————, *Juliette*, ed. Austryn Wainhouse (New York: Grove Press, 1968).

Diderot, Denis, *Œuvres philosophiques*, ed. Paul Vernière (Paris: Garnier, 1964).

[Faselius, Johann Friedrich,] *Elements of Medical Jurisprudence*, tr. Samuel Farr (London, 1788).

Goldsmith, Oliver, *History of the Earth and Animated Nature*, 8 vols (2nd ed., London, 1779).

Hume, David, *Commentaries on the Law of Scotland, Respecting Crimes*, 2 vols (2nd ed., Edinburgh, 1819).

Hunter, John, *Observations on Certain Parts of the Animal Oeconomy* (2nd ed., London, 1792).

Jefferson, Thomas, *Notes on the State of Virginia* (2nd American ed., Philadelphia, 1794).

Kemeys, John Gardner, *Free and Candid Reflections Occasioned by the Late Additional Duties on Sugars and on Rum* (London, 1783).

Lawrence, William, *Lectures on Physiology, Zoology, and the Natural History of Man* (London, 1819).

[Long, Edward], *Candid Reflections Upon … The Negroe-Cause* (London, 1772).

————, *The History of Jamaica*, 3 vols (London, 1774).

Mackenzie, Sir George, *The Laws and Customes of Scotland, in Matters Criminal* (Edinburgh, 1678).

Ogilby, John, *Africa* (London, 1670).

Old Bailey Proceedings Online (www.oldbaileyonline.org, version 7.2, 8 January 2016).

Pallas, Peter Simon Neue Nordische Beyträge, 1 (1781), 156–57.

Parsons, James, *A Mechanical and Critical Enquiry into the Nature of Hermaphrodites* (London, 1741).

Prichard, James Cowles, *Researches into the Physical History of Man*, ed. George W. Stocking, Jr. (Chicago: University of Chicago Press, 1973 [1813]).

Ray, John, *The Wisdom of God Manifested in the Works of the Creation* (London, 1691).

'Rousseau, Jean Jaques, Citoyen de Genève', *Discours sur l'origine et les fondemens de l'inegalité parmi les hommes* (Amsterdam, 1755).

[Smalbroke, Richard], *Reformation Necessary to Prevent Our Ruine* (London, 1728).

[Stretser, Thomas], *A New Description of Merryland* (4th ed., Bath, 1741).

Swift, Jonathan, *Gulliver's Travels*, ed. Claude Rawson (Oxford: Oxford University Press, 2005).

Topsell, Edward, *The Historie of Foure-Footed Beasts* (London, 1607).

Tyson, Edward, *Orang-Outang, sive Homo Sylvestris: Or, the Anatomy of a Pygmie* (London, 1699).

Voltaire, *Candide and Other Stories*, tr. Roger Pearson (Oxford: Oxford University Press, 1990).

Zimmermann, E. A. W. Von, *Geographische Geschichte des Menschen*, 3 vols (Leipzig, 1778–83).

Secondary sources

Attwood, Feona, 'Reading Porn: The Paradigm Shift in Pornography Research', *Sexualities*, 5 (2002), 91–105.

Beetz, Andrea M., and Anthony L. Podberscek (eds), *Bestiality and Zoophilia: Sexual Relations with Animals*, special issue of *Anthrozöos* (2005).

Ben-Atar, Doron S., and Richard D. Brown, *Taming Lust: Crimes Against Nature in the Early Republic* (Philadelphia: University of Pennsylvania Press, 2014).

Betteridge, Tom (ed.), *Sodomy in Early Modern Europe* (Manchester: Manchester University Press, 2002).

Bierne, Piers, 'Rethinking Bestiality: Towards a Concept of Interspecies Sexual Assault', *Theoretical Criminology*, 1 (1997), 317–40.

Boer, Roland, 'From Horse Kissing to Beastly Emissions: Paraphilias in the Ancient Near East', in Mark Masterson, Nancy Sorkin Rabinowitz, and James Robson (eds), *Sex in Antiquity: Exploring Gender and Sexuality in the Ancient World* (London: Routledge, 2015), 67–79.

Creager, Angela N. H., and William Chester Jordan (eds), *The Animal-Human Boundary: Historical Perspectives* (Rochester: University of Rochester Press, 2002).

Cressy, David, *Agnes Bowker's Cat: Travesties and Transgressions in Tudor and Stuart England* (Oxford: Oxford University Press, 2000).

Daston, Lorraine, and Katharine Park, *Wonders and the Order of Nature* (New York: Zone, 2001).

Dawkins, Marian, 'Animal Welfare and the Paradox of Animal Consciousness', *Advances in the Study of Behavior*, 47 (2015), 5–38.

Dekkers, Midas, *Dearest Pet: On Bestiality*, tr. Paul Vincent (London: Verso, 2000).

Douthwaite, Julia V., *The Wild Girl, Natural Man, and the Monster: Dangerous Experiments in the Age of Enlightenment* (Chicago: University of Chicago Press, 2002).

Earls, Irene, *Renaissance Art: A Topical Dictionary* (Westport: Greenwood Press, 1987).

Feuerstein-Herz, Petra, 'Eberhard August Wilhelm von Zimmermann (1743–1815) und die Tiergeographie', Diss. (Technische Universität Braunschweig, 2003).

Fox, Christopher, Roy Porter, and Robert Wokler (eds), *Inventing Human Science: Eighteenth-Century Domains* (Berkeley: University of California Press, 1995).

Fudge, Erica, 'Monstrous Acts: Bestiality in Early Modern England', *History Today*, 50/8 (2000), 20–25.

————, *Perceiving Animals: Humans and Beasts in Early Modern English Culture* (Basingstoke: Macmillan, 2000).

Fuss, Diana (ed.), *Human, All Too Human* (New York: Routledge, 1996).

Hoffner, Harry A., Jr., 'Hittite Laws', in *Law Collections from Mesopotamia and Asia Minor*, ed. Martha T. Roth (Atlanta: Scholars Press, 1995).

Hund, Wulf D., Charles W. Mills., and Silvia Sebastiani (eds), *Simianization: Apes, Gender, Class, and Race* (Racism Analysis Yearbook, 6: 2015/16; Zürich: Lit, 2015).

Janson, H. W., *Apes and Ape Lore in the Middle Ages and the Renaissance* (London: Warburg Institute, 1952).

Jolivet, Vincent, 'Lumières et bestialité', *Dix-huitième siècle*, 42 (2010), 285–303.

Kinsey, Alfred C., Wardell B. Pomeroy., Clyde E. Martin., and Paul H. Gebhard., *Sexual Behavior in the Human Female* (Philadelphia: W. B. Saunders, 1953).

Liliequist, Jonas, 'Peasants against Nature: Crossing the Boundaries between Human and Animal in Seventeenth- and Eighteenth-Century Sweden', *Journal of the History of Sexuality*, 1 (1991), 393–423.

Masterson, Mark, Nancy Sorkin Rabinowitz, and James Robson (eds), *Sex in Antiquity: Exploring Gender and Sexuality in the Ancient World* (London: Routledge, 2015).

Maxwell-Stuart, P. G., '"Wild, filthie, execrabill, detestabill, and unnatural sin": Bestiality in Early Modern Scotland', in Tom Betteridge (ed.), *Sodomy in Early Modern Europe* (Manchester: Manchester University Press, 2002), 82–93.

Meijer, Miriam Claude, *Race and Aesthetics in the Anthropology of Petrus Camper (1722–1789)* (Amsterdam: Rodopi, 1999).

————, 'Bones, Law, and Order in Amsterdam: Petrus Camper's Morphological Insights', in Klass Van Berkel and Bart Ramakers (eds), *Petrus Camper in Context: Science, The Arts, and Society in the Eighteenth-Century Dutch Republic* (Hilversum: Verloren, 2015), 187–213.

Miletski, Hani, *Understanding Bestiality and Zoophilia* (Bethseda, MD: East-West Publishing, 2002).

Murrin, John M., '"Things Fearful to Name": Bestiality in Early America', in Angela N. H. Creager and William Chester Jordan (eds), *The Animal-Human Boundary: Historical Perspectives* (Rochester: University of Rochester Press, 2002), 115–56.

Niekerk, Carl, 'Man and Orangutan in Eighteenth-Century Thinking: Retracing the Early History of Dutch and German Anthropology', *Monatshefte für Deutschsprachige Literatur und Kultur*, 96 (2004), 477–502.

Panksepp, Jaak, 'The Basic Emotional Circuits of Mammalian Brains: Do Animals Have Affective Lives?', *Neuroscience and Behavioral Reviews*, 35 (2011), 1791–1804.

Parker, Graham, 'Is a Duck an Animal? An Exploration of Bestiality as a Crime', *Criminal Justice History*, 7 (1986), 95–109.

Ritvo, Harriet, 'Barring the Cross: Miscegenation and Purity in Eighteenth- and Nineteenth-Century Britain', in Diana Fuss (ed.), *Human, All Too Human* (New York: Routledge, 1996), 37–57.

Rossiianov, Kirill, 'Beyond Species: Il'ya Ivanov and His Experiments on Cross-Breeding Humans with Anthropoid Apes', *Science in Context*, 15 (2002), 277–316.

Rydström, Jens, *Sinners and Citizens: Bestiality and Homosexuality in Sweden, 1880–1950* (Chicago: University of Chicago Press, 2003).

Salih, Sara, 'Filling Up the Space between Mankind and Ape: Racism, Speciesism and the Androphilic Ape', *Ariel: A Review of International English Literature*, 38 (2007), 95–111.

Sebastiani, Silvia, 'Challenging Boundaries: Apes and Savages in Enlightenment', in Wulf D. Hund, Charles W. Mills, and Silvia Sebastiani (eds), *Simianization: Apes, Gender, Class, and Race*, 105–37.

Sloan, Philip, 'The Gaze of Natural History', in Christopher Fox, Roy Porter, and Robert Wokler (eds), *Inventing Human Science: Eighteenth-Century Domains* (Berkeley: University of California Press, 1995), 112–51.

Sommerville, Margaret R., *Sex and Subjection: Attitudes to Women in Early Modern Society* (London: Arnold, 1995).

Tague, Ingrid H., *Animal Companions: Pets and Social Change in Eighteenth-Century Britain* (University Park: State University of Pennsylvania Press, 2015).

Thomas, Keith, *Man and the Natural World* (London: Penguin, 1984).

Thomas, Courtney, '"Not Having God before His Eyes": Bestiality in Early Modern England', *Sixteenth Century*, 26 (2011), 149–73.

Toulalan, Sarah, and Kate Fisher (eds), *The Routledge History of Sex and the Body, 1500 to the Present* (London: Routledge, 2013).

Tuttle, Russell H., *Apes and Human Evolution* (Cambridge: Harvard University Press, 2014).

Urness, Carol (ed.), *A Naturalist in Russia: Letters from Peter Simon Pallas to Thomas Pennant* (Minneapolis: University of Minnesota Press, 1967).

Van Berkel, Klaas, and Bart Ramakers (eds), *Petrus Camper in Context: Science, The Arts, and Society in the Eighteenth-Century Dutch Republic* (Hilversum: Verloren, 2015).

Wiesner, Merry E., *Women and Gender in Early Modern Europe* (2nd ed., Cambridge: Cambridge University Press, 2000).

7 'A disgusting exhibition of brutality'

Animals, the law, and the Warwick lion fight of 1825

Helen Cowie

On 26 July 1825, the usually quiet market town of Warwick was host to an extraordinary spectacle: a fight between a lion and six bulldogs. The event was organised by the showman George Wombwell, owner of a popular travelling menagerie. The lion was a handsome beast called Nero, born in captivity and widely esteemed 'a beautiful and majestic animal'; the dogs were veteran fighters, known for their strength and ferocity.[1] Around 500 people congregated to watch the combat, which, against expectations, ended in victory for the dogs.

The Warwick lion fight attracted considerable attention in the contemporary press, most of it sharply critical. Fights between exotic animals had once been commonplace at royal courts; the last contest between a lion and three mastiffs was held in 1610, 'in the presence of James I and his son Prince Henry'.[2] By the beginning of the nineteenth century, however, attitudes towards animal baiting were starting to change, and what had once been an accepted source of courtly entertainment was no longer viewed with equanimity. In this new social climate, the spectacle at Warwick appeared to many as brutal, morally repugnant and anachronistic, a remnant of barbarism in a modernising age.

This chapter offers a detailed study of the Warwick lion fight, its reception and consequences. Noteworthy for its drama and novelty, the lion bait was particularly significant due to its timing, falling during an important decade for human–animal relations. When Wombwell first announced the forthcoming contest, the rights of animals were becoming a matter for discussion. The House of Commons was debating a bill to ban bull and bear baiting, magistrates across Britain were grappling with the interpretation of existing laws against animal cruelty, and the newly formed Society for the Prevention of Cruelty to Animals was campaigning to stop a variety of abuses, from whipping pigs to death to boiling lobsters alive.[3] Under these circumstances, the lion fight assumed new importance, generating much debate in newspapers and periodicals. Through analysis of this material, we can learn much about contemporary attitudes towards animals in a period when their treatment was closely connected with issues of class, public decency, and national identity.

From cruelty to compassion?

The early nineteenth century has traditionally been seen as an era of increased compassion for animals. In previous centuries, a culture of sympathy for animals was lacking, in Britain as elsewhere in Europe. Around 1800, however, a shift in attitudes started to take place and resulted in greater concern for animal welfare (albeit, as we shall see, of a rather selective nature), giving rise to the modern 'connection between Englishness and kindness to animals'.[4]

Historians have offered a variety of explanations for this change. Keith Thomas argues that it was the culmination of a gradual shift in attitudes towards non-humans. These grew, in part, out of new religious principles – most often held by non-conformists such as Quakers and Methodists – which stressed that it was man's duty to take care of God's creation. They also grew out of a more secular emphasis on 'sensation and feeling as the true basis for a claim to moral consideration'[5] – a mentality best encapsulated in Jeremy Bentham's famous dictum, that it is not animals' capacity to 'reason' or to 'talk' that entitles them to humane treatment, but their capacity to 'suffer'.[6] Both of these lines of thinking also played a key role in the rise of abolitionism, which likewise came to prominence in the late eighteenth century.

At the same time as new ideas influenced the thinking of a (probably small)[7] minority of the upper classes, major social changes further affected the way people treated animals. James Turner suggests that sensitivity to animal welfare was heightened by the social re-structuring resulting from industrialisation and urbanisation. On the one hand, the new work demands imposed by factory discipline supplanted the more uneven rhythms of the workshop and made drawn-out and violent blood sports like bull baiting appear socially corrosive and archaic. On the other, a sense of nostalgia for the natural world arose among the newly urbanised middle classes, stimulating feelings of compassion towards the brute creation – especially farm animals, reminiscent of a rural idyll.[8] Hilda Kean and Diana Donald, challenging this last point, argue that it was the continued presence of farm animals in cities and their critical role in the economy, rather than their absence from city life, that fostered concerns for animal welfare, since the sight of visible cruelty on the streets shocked and upset middle-class viewers: 'It was not philosophical distance from sites of cruelty, but painful proximity to them which prompted Londoners' protests'.[9] Either way, there was a fear that perpetrating or even just witnessing acts of cruelty to animals would brutalise the masses, triggering moral degeneration and social disorder. This concern was encapsulated in William Hogarth's famous *Four Stages of Cruelty* (1751), in which a young man named Tom Nero begins his descent into brutality by butchering dogs and beating a horse and ends up murdering his pregnant lover.[10]

Increasing compassion for animals was reflected in the introduction of legislation to prohibit some of the most blatant forms of cruelty, particularly those closely associated with social unrest. From the early eighteenth century, individual towns banned the practice of cock-throwing, a brutal fairground entertainment in which a live cock was tethered to a pole and bombarded with missiles. This was deemed particularly reprehensible because of its gross unfairness (a point that would also be raised about the Warwick lion fight) and may also have been targeted because it was predominantly a plebeian sport. By the end of the century, the remit of reformers had widened and bull and bear baiting were starting to come in for criticism. In 1800, MP William Pulteney introduced a bill into Parliament to abolish bull baiting. This bill failed, thwarted by opponent William Windham's claims that it placed an unfair burden on the poor, who would soon be 'despoiled of all their enjoyments'.[11] However, after two further failures (in 1802 and 1809), Richard Martin (MP for Galway) finally succeeded in pushing a bill through Parliament making it a crime to mistreat any farm or draught animal. Though the 1822 Animal Cruelty Act did not explicitly outlaw bull and bear baiting or any other blood sports, it was interpreted by some magistrates as doing so, forming the basis for future animal cruelty legislation.

To police the treatment of animals, the Society for the Prevention of Cruelty to Animals (SPCA) was founded in 1824 by Reverend Arthur Broome. The Society was originally centred in London, but it quickly began to acquire affiliated branches in the provinces. The SPCA recruited a team of uniformed inspectors to apprehend perpetrators of abuse and collaborated actively with the authorities in the suppression of animal cruelty. In the first year of its existence it secured 144 convictions for cruel and improper treatment of animals, 'chiefly among bullock drivers, but partly also among those who have the care of horses, and who by furious driving or riding, not only injure these animals, but incur great hazard of destruction of human life'.[12] The Society attempted simultaneously to reform public morals, arranging for 'the circulation of suitable tracts gratuitously ... among persons intrusted [sic] with cattle' and encouraging 'the introduction into schools of books calculated to humanize the mind of youth'.[13] Though in many ways a moderate organisation, the SPCA (RSPCA from 1840, when it gained royal patronage) continually tested the limits of the existing legislation on animal cruelty, pushing, at different times, for an end to blood sports, for improvements to knackers' yards, for the regulation of vivisection, and for the extension of the laws to include wild as well as domestic animals. In the mid-1820s, the SPCA's primary focus was the abolition of bull baiting and bear baiting, pursuits it regarded as both inimical to animal welfare and detrimental to human morality.

The fight

Against this political and social backdrop the first notice of the proposed lion combat appeared. The fight was advertised in several provincial papers, beginning with *Berrow's Worcester Journal* in March 1825. A notice in this publication stated that the combat would take place at Worcester race course and see Wombwell's famous lion, Nero, take on 'six mastiffs' for a wager of 5,000 sovereigns. The contest would occur in a specially constructed 'circular den,

36 feet in circumference'. The lion, according to reports, was a phenomenal animal, aged six years old and standing '4 ½ feet high' and '13 feet in length'.[14]

By April, when no fight had happened, doubts started to surface as to the veracity of the story. The journal *Bell's Life in London* concluded that the whole thing had been a hoax, fabricated by Wombwell to drum up trade for his menagerie. It noted that no accounts of the fight had appeared in the Birmingham papers, which 'would scarcely have been silent on so interesting an occurrence' and expressed doubts as to the reported dimensions of the lion, certain there had never been 'a lion in existence of such magnitude, thirteen feet being about the length of two prize oxen together, and completely out of proportion with the alleged height'.[15] These conclusions were reprinted in several national and local papers and the lion fight dismissed as a fraud.

Despite this inauspicious start, rumours persisted that a lion fight was imminent, and towards the end of June a new location was reported – Warwick. There was scepticism as to the truth of this information, given the earlier non-event at Worcester, but the story gathered momentum when the *Hereford Independent* reported that a 'den or cage' manufactured in Shrewsbury for 'the combat between the lion Nero and the mastiffs' was en route to Northampton.[16] The re-arranged fight was to take place on Tuesday 26 July in the yard of a disused factory.

The build-up for the fight began in earnest on Monday 18 July, when Wombwell arrived in Warwick with his menagerie. The caravans were stationed in the factory grounds and positioned to form 'a compact square' within which the fight would take place. A special cage was dragged into the centre, 'the bars sufficiently far asunder to permit the dogs [but not the lion!] to pass in and out', and terraced seating was erected on top of the other caravans to accommodate spectators.[17] In the days before the fight, Wombwell paraded through the town with a camel 'for the diversion of the idlers'. The menagerie was opened to the public as usual, 'drawing the country folk pretty freely for a sight of the lion'.[18]

On Friday 22 July, the dogs chosen to engage the lion were brought up from London and exhibited at the Green Dragon Inn in Warwick, where 'a great number of persons paid sixpence each to have an opportunity of judging their qualities'.[19] There were eight dogs in total, from which six were selected to fight Nero. According to *Jackson's Oxford Journal*, the animals chosen for the combat were 'good-looking savage vermin, averaging about 40lbs in weight'. The *Caledonian Mercury* named the canine combatants as

> 1) Turk, a brown-coloured dog; 2) Captain, a fallow and white dog, with a skewbald face; 3) Tiger, a brown dog with white legs; 4) Nettle, a little brindled bitch, with black head; 5) Rose, a brindle-pied bitch; [and] 6) Nelson, a white dog with brindled spots.[20]

Another dog, weighing 'over 70lbs' and 'of most villainous aspect', also formed part of the contingent, but Wombwell refused to allow this specimen to engage his lion, fearing that it might cause Nero serious injury.[21] The canines were cared for by Mr Edwards, John Jones, William Davis, and Samuel Wedgebury, a well-known London dog-fancier.[22]

As the appointed time of the fight approached, Wombwell took additional steps to publicise the contest. On Tuesday morning, the showman permitted 'several persons' to enter the factory yard in order to inspect preparations. Around 10 am, the dogs themselves were brought into the menagerie and introduced to Nero, who reportedly 'eyed them with great complacency'.[23] Later in the afternoon Wombwell dispatched trumpeters into the surrounding villages and the neighbouring town of Leamington Spa to 'announce the fight'.[24] The latter succeeded in enticing some people to Warwick, but not as many as hoped: partly because the majority of prospective viewers believed that the whole thing was a hoax and partly because Wombwell charged exorbitant fees for attending. According to *Jackson's Oxford Journal*, prices ranged from 3 guineas 'for seats at the windows in the first, second and third floors of the unoccupied manufactory' to 'half a guinea for standing room in the square', confining the audience to 'gentlemen resident in the neighbourhood'.[25]

As Wombwell put the finishing touches to his spectacle, he received an unexpected visit from a Quaker named Samuel Hoare, who entreated the showman to cancel the fight. Addressing Wombwell in a letter, Hoare appealed to Wombwell's humanity, asking, 'how thou wilt feel to see the noble animal thou hast so long protected, and which has been in part the means of supplying thee with the means of life, mangled and bleeding before thee?' The Quaker condemned the menagerist for allowing 'this cruel and very disgraceful exhibition' to go ahead and exhorted him to consider the probable consequences for his immortal soul; as he put it, 'He who gave life did not give it to be the sport of cruel man' and 'will certainly call man to account for his conduct towards dumb creatures'. Indeed, the animals' muteness was a reason for according them special consideration, rather than a justification for harming them, for such conduct directly violated God's will: 'It is unmanly, it is mean and cowardly, to torment anything that cannot defend itself, that cannot speak to tell its pains and sufferings, that cannot ask for mercy'. These eloquent arguments resonated with some contemporary journalists, who published the letter in full in their respective newspapers. They appear, however, to have had little impact on Wombwell, who refused to abandon the contest.[26]

By five in the evening preparations were complete, and the first spectators began to arrive. Around 500 assembled in total – fewer than Wombwell had anticipated.[27] Of these, about half crowded into the yard, while the rest positioned themselves in the factory windows and above the caravans. Most of those present occupied the seats 'which were most distant from the lion, from an apprehension that the unfortunate animal might break forth and take revenge on his tormentors'.[28]

Once the spectators were in place, Nero was led from his travelling van to the fighting cage, and the rules of the combat were read out. The contest was to consist of three rounds, with three dogs being set on Nero at a time. If either Nero or any of the dogs ran away, they would be considered beaten and forced to withdraw. Should Nero prevail against the first three dogs he would be given a twenty-minute respite in which to recover before the second three were introduced into his den.[29]

At around ten to seven, Wombwell, who had been in the cage with Nero, left the arena and the first set of dogs entered through the railings. Nero showed no desire to attack them, walking quietly around the cage and at length lying down in one of the corners 'with his nose through the bars'. The dogs, however, immediately assailed the lion, 'darting with fury upon him' and snapping at his mouth and nostrils. 'They unceasingly kept goading him, biting and darting at his nose, sometimes hanging from his mouth, or one of them endeavouring to pin a paw while the others mangled his head'.[30] Though Nero eventually managed to maim one of the dogs, it was several minutes before he could rid himself of the other two, Tiger and Turk. According to eyewitness accounts, Nero never once exerted his full strength against the dogs, swatting them away with his paws rather than making use of his teeth. As the *Examiner* expressed it, 'from the beginning of the matter to the end the lion was merely a sufferer – he never struck a blow'.[31]

As agreed in advance, Nero was accorded a twenty-minute respite before the second set of dogs was unleashed upon him. During this interval, Wombwell went back into the cage and gave the lion a pan of water. Nero lapped it up with gusto, and Wombwell sprinkled the remainder on his face to refresh him. While this was happening, a 'crowd of rustics' outside the factory yard attempted to break into the exhibition, some of them scaling the temporary walls around the arena.[32] There was consternation amongst the spectators, who feared an intrusion from the mob, but the incident was quickly dealt with by Wombwell, who set about clearing the ground 'with the aid of a pretty good stick'.[33] By the time the intruders had been removed, three-quarters of an hour had elapsed, giving Nero additional time to recuperate.

With the crowd growing increasingly impatient, the second round of the contest finally began. A further three dogs, Nettle, Rose, and Nelson, were slipped into Nero's cage, and immediately ran towards him, bringing him down and pinning him to the floor. Nero roared mournfully and tried to escape, but was soon immobilised once more. When he failed to free himself, the owners of the dogs stepped in to remove their animals, calling on Wombwell to surrender on behalf of his lion. Initially the showman refused to concede defeat, insisting that a third round take place, but when the dogs again got the better of Nero Wombwell gave in, fearful that 'the death of the animal must be the consequence of further punishment'. The lion fight had lasted for just over an hour, a mere sixteen minutes of which was taken up by actual combat.[34]

The reprise

Despite the disappointment of the first lion fight, Wombwell consented to a second contest less than a week later in a bid to redeem the reputation of his lions. The main combatant on this occasion was the Scottish-born lion, Wallace, known for his fiery temper. The dogs, again supplied by Wedgebury, were Tinker, Sweep, Ball, Turpin, Tiger, and the 'canine hero' Billy, famous for killing 100 rats in nine minutes.[35] The rules of the contest were similar to those of the first fight, with the exception that the dogs would face Wallace in pairs rather than three at a time.

While the format of the fight was almost identical, the outcome was very different. Though born in captivity like Nero, Wallace was a ferocious animal, so unpredictable that 'it is but seldom he lets even his feeders approach him'. Faced with the first couple of dogs, the lion immediately attacked them, holding one animal in his teeth and 'deliberately walk[ing] around the stage with him as a cat would a mouse'. The other dog was badly wounded from a stroke of Wallace's paw and 'died just a few seconds after he was taken out of the cage'.

In the second and third rounds, Wallace continued to dominate, assailing his canine opponents until they fled or were rescued by their owners. One dog, the rat-catcher, Billy, remained clamped in the lion's jaws until a keeper 'threw a piece of raw flesh into the den' to tempt him away. Another dog was left in a critical condition with 'several of his ribs broken'.[36] According to a spectator, Wallace conquered the dogs so easily that 'he would have eaten his way to London through them, though standing as thick together as a mob at the elections'.[37] A subsequent newspaper report painted a gruesome picture of the scene inside the arena, describing how Wallace strode malevolently around the cage, his jaws 'covered with crimson foam' and his feet 'printing each step with gore'.[38]

The Fight between the Lion WALLACE & the dogs TINKER & BALL, in the Factory Yard, in the Town of Warwick.

Figure 7.1 'The Fight between the Lion Wallace and the Dogs Tinker and Ball in the Factory Yard in the Town of Warwick', from *Anecdotes, Original and Selected, Of The Turf, The Chase, The Ring, And The Stage* by Pierce Egan, published 1825 (etching), Lane, Theodore (1800–28) / Private Collection / The Stapleton Collection / Bridgeman Images.

Theodore Lane's cartoon of the fight (Figure 7.1) captures the savagery of the contest and gives us a sense of how events transpired. Wallace stands in the centre of the cage, grasping one blood-spattered dog between his jaws and pinning another with his paw. Wombwell (in the black hat) crouches on the left of the stage, urging his lion on, while the two dog handlers, Wedgebury and Edwards, watch on from the other side. The assembled crowd, from their attire of top hats and frock coats, appear to be predominantly members of the gentry, as suggested in contemporary reports. Some of the figures watching the scene from the windows on the right may possibly be women; according to one report, several ladies 'witnessed the exhibition from the upper apartment of the factory', enjoying the spectacle from a safe distance.[39]

The response

The Warwick lion fights received extensive coverage in the local and national papers, much of it deeply unfavourable. Reported at length in several contemporary journals, the combats were criticised for their unequal nature and evident cruelty. Commentators found the violence of the fights distasteful and condemned the organisers for allowing them to go ahead. They also reflected on the wider implications of blood sports for human society, drawing a close connection between violence towards animals and violence towards humans. The content of these discussions gives us an insight into broader attitudes towards animals in the mid-1820s, a period when bull and bear baiting were hot topics in Parliament and in which the first legislation to protect non-human subjects was being introduced into Britain.

First, coverage of the first lion fight emphasised the brutality of the contest. Newspaper reports recounted the event in graphic language, dwelling upon its more lurid details and expressing sympathy for Nero. *Jackson's Oxford Journal* described Nero's pain during the combat, reporting that he 'roared with anguish' when the dogs were set upon him for a second time and was later to be seen 'panting on the stage, his mouth nose and chops full of blood'.[40] Radical journalist William Hone singled out eighteen separate 'points of cruelty' in his later summary of the fight, all of which evoked pity for Nero. In point five, the beleaguered lion was described as turning 'a forlorn and despairing' glance towards his keepers as the dogs assaulted him for a second time, as if begging 'for assistance'. Point twelve detailed how the dog Turk 'flew at [Nero's] nose and there fastened himself like a leech' while the lion bellowed with pain.[41]

Many of the papers humanised Nero, emphasising his placid, gentle nature. The *Caledonian Mercury* described him at one point as 'docile as a spaniel', while the *Examiner*, employing a less flattering cross-species analogy, claimed that Nero 'had no more thought or knowledge of fighting than a sheep would have had under the same circumstances'.[42] Nero's refusal to defend himself detracted from the legitimacy of the contest, underlining its essential injustice. Similar language was employed to describe other animals reputedly forced to fight against their will, among them a young bear at the

Holborn pit, which, though 'much lacerated' and 'un-muzzled', evinced 'the utmost docility and good nature, holding up its bleeding paw as if to reproach his cruel master'.[43]

Of all the papers, the most vocal in its denunciation of the combat was the *Liverpool Mercury*, which published a highly critical commentary of the fight three days after it occurred. In this damning article, the *Mercury* stigmatised the brutal contest as 'a disgraceful blot on the national character'. Had the fight had some scientific value, it might just have been permissible, for maybe there was something to be learned about the relative strength of the lion and the dog. As things were, however, no dazzling zoological insights could be expected to emerge from a contest that pitted Nero, 'born in a cage in which he can scarcely turn himself, labouring under the effects of long and unnatural restraint and diet', against 'a pack of fierce dogs in the finest possible condition, both as to training and nourishment'; all that was likely to be gratified was a love of sensation.[44]

Beyond the abuse of the lion, the *Mercury* also took particular issue with the 'brutalising' effect the fight was likely to have on spectators and organisers, who might easily graduate from torturing defenceless beasts to killing fellow human beings. Even before the contest took place, the *Mercury* branded it 'a disgrace ... to our country, and to the age in which we live'.[45] As precise details of the fight filtered out, the paper (whose editor had some fifteen years previously attempted to form a society for the protection of animals in Liverpool) grew even more vehement in its criticisms, describing the incident as symptomatic of 'that ferocious and unchristian spirit which appears to be alarmingly on the increase in this country'. To underline the point, the report concluded with a pointed reference to Hogarth's 'Four Stages of Cruelty', in which, as we have seen, the protagonist 'begins with tormenting dogs, cats and inferior animals and ends his career by the murder of a fellow creature'.[46] Such reflections echoed the comments of one opponent of bull-baiting, who identified a direct connection between these 'spectacles of cruelty and blood' and 'the numerous instances of atrocious murders which have occurred during these last few years'.[47] While some commentators pondered the links between cruelty to animals and human-on-human violence, another theme that emerged strongly from contemporary reports was contempt for Wombwell for deliberately exposing Nero to harm. The lion, it was generally accepted, was a majestic but benign animal with no inclination to fight. Wombwell, however, purely for profit had agreed to let him face a pack of highly trained dogs, wantonly risking his life and putting monetary gain above his duty towards the animal. This was doubly culpable because Wombwell had reared Nero from a cub and should have acted as his protector. As the *Times* expressed it:

> it would be a joke to say anything about the feelings of any man who, for the sake of pecuniary advantage, could make up his mind to expose a noble animal which he had bred, and which had become attached to him, to a horrible and lingering death.[48]

Royal Menagerie. Exeter Change. Strand.

Figure 7.2 Robertson's Royal Menagerie, Exeter Change, Strand (coloured engraving), English School (19th century) / Private Collection / © Look and Learn / Peter Jackson Collection / Bridgeman Images.

The semi-domestic nature of the lion, and its dependence upon its human owner, complicated the creature's status, blurring the boundary between wild animal and treasured pet.[49] As captive beasts, menagerie animals like Nero socialised extensively with humans and often enjoyed quite intimate relationships with their keepers (Figure 7.2). Wombwell, for instance, recounted his close personal contact with all of the animals in his menagerie, which ranged from paring the hooves of a 'vicious' zebra to giving the elephant his bucket of 'strong ale' and rolling his snakes in blankets each night to keep them warm.[50] A contributor to *Blackwood's Magazine* in 1855, meanwhile, recalled his childhood experiences with Nero, 'into whose cell you might go...for the moderate fee of half a crown', and Wallace, 'a rampant, reddish-maned animal...[who] would not tolerate the affront of being roused by the application of a long pole'.[51] These kinds of interactions elevated some beasts to the level of quasi-pets, with names, specific needs, and in the case of the two lions, individual personalities. To mistreat an animal that had been raised in this way appeared tantamount to a betrayal.

Finally, on a more prosaic level, the Warwick lion fight was condemned in the press as a threat to public order and a source of national disgrace. Like other contemporary blood sports, the fight was seen as an invitation to gambling and public immorality. Though, in the event, Wombwell's high prices kept the lower classes out of the arena, their attempted intrusion between the first and second rounds highlighted the potential for disorder and raised the spectre of mob violence.

At a time when public animal baiting was becoming a matter for debate, moreover, it was suggested that the lion fight was anachronistic and un-British. The *Liverpool Mercury*, as we have seen, described the contest as 'a disgraceful blot on the national character'.[52] William Hone made a similar point when he equated the fight with 'the gladiatorial shows of Rome, the quail fights of India' and 'the bull-fights of Spain'.[53] These explicit allusions to besmirched national honour and foreign barbarity revealed a growing sense that cruelty to animals was a sign of cultural regression and a belief that certain blood sports, rather than nurturing manly courage, precipitated crime and paved the way for political oppression. Speaking in a debate on bull-baiting earlier in 1825, for instance, MP William Smith made a direct connection between Spanish bull-fighting and the barbarities of the Inquisition, theorising that a 'people accustomed to see the wanton shedding of the blood of animals' had 'been prepared to witness the shedding of human blood at the *auto da fé*'.[54] Fifty years later, as the British warmed to their burgeoning reputation as lovers of animals, the RSPCA's monthly magazine *The Animal World* conjured similar national stereotypes, while at the same time warning against complacency at home:

> It is the custom in England when reference is made to the torturing of animals for providing pleasure to human beings to shrug one's shoulders, sigh, think of Spain and perhaps mutter something about the debasement of a people whose national pleasure is the bull fight.[55]

Cruelty to animals thus became a sign of national depravity and a source of communal shame. As one commentator remarked in 1838,

> what dogs and lions can achieve in the arena of combat … having now been ascertained, let us hope that no closer approximation to the sanguinary games of the Roman amphitheatre may ever be attempted in Great Britain, nor her soil again polluted by a repetition of such spectacles.[56]

Legal loopholes

While the increasingly vocal periodical press expressed its disgust at the Warwick lion fight, the contest also had repercussions in two other arenas: the House of Commons and the law courts. In the former, the lion bait coincided with a series of debates taking place on the subject of bull and bear baiting. In the latter, it raised important questions about the scope of existing animal cruelty legislation, and whether it included dogs and lions.

Animal baiting was already a contentious issue in Parliament before the Warwick lion fight took place. Early attempts to introduce animal cruelty legislation had focused on bull- and bear-baiting, which were seen as socially corrosive as well as inhumane. Martin's Act of 1822, however, only explicitly applied to the treatment of horses and cattle (though, as we shall see, its terms were open to interpretation), and efforts were made throughout the 1820s to extend its coverage into new areas. In March 1825, for example, when the first rumours of the lion fight were in the air, Martin introduced a bill to outlaw bear-baiting, cock-fighting, and dog-fighting.[57] In February 1826, he made a second attempt to enact the measure, suffering defeat in the Commons by a majority of 39.[58] The key issues at stake in these debates were whether additional legislation was necessary (some thought the existing laws were adequate), whether the blood sports covered were indeed cruel (one MP claimed that the bear used in the Westminster pit was in good health and had recently been retired from baits, not because of injury, but because 'he had grown too fat for the exercise') and whether the proposed measures were socially just.[59] Those in favour of a ban on blood sports emphasised the 'gross and wanton cruelty' of such pastimes and their capacity to corrupt both public morals and animal bodies; the veterinary surgeon William Youatt even put forward a theory that training dogs to fight increased their susceptibility to hydrophobia (rabies), a much-feared disease in nineteenth-century Britain.[60] Those who opposed the ban tended to highlight its social bias, noting that it would suppress the sports of the lower classes while allowing the elite to continue hunting foxes and shooting pigeons. Robert Peel, one of the most vocal critics of the bill, thus objected that it would 'debar the lower classes of society from those amusements which persons of rank and station were to continue in the enjoyment of', allowing a magistrate 'after having dined upon crimped cod and ... devoted the whole of his day to fox-hunting' to punish another man for baiting a bull.[61]

The Warwick lion fight was held just as these discussions were going on and appears to have played a part in their outcome. Though the more mundane pastimes of bull- and bear-baiting received most publicity, the notoriety of the lion bait and its widespread coverage attracted significant attention, pushing some to vote in favour of Martin's bill. One MP, Colonel Wood, cited the 'brutal, cowardly and atrocious scene [that] was permitted to take place at Warwick' as his main reason for supporting the legislation, which he had previously regarded as unnecessary. Wood had not seen the fight for himself, but had read an account of it (probably one of the newspaper articles cited above), from which he quoted liberally in the Commons. He concluded that 'the noble animal which had been so tormented was more worthy of protection than its brutal owner' and that 'some law ought to have sufficient force to prevent a similar occurrence'.[62] The lion fight therefore influenced contemporary perceptions of blood sports by putting them centre stage in the public consciousness and showing the distasteful extremes to which showmen might go if not constrained by the law. The unusual nature of the combatants may also have shaped the wording of the successful Animal Cruelty Act of 1835, which was careful to outlaw 'baiting or running bulls, or baiting or fighting bears, dogs or other animals (*whether domestic or wild nature or kind*) or fighting cocks'.[63]

While Parliament debated potential extensions and additions to law in regards to animals, magistrates and judges attempted to enforce the existing legislation, which itself remained a matter for discussion. The original Act of 1822 officially made it an offence to 'wantonly and cruelly beat, abuse, or ill-treat any Horse, Mare, Gelding, Mule, Ass, Ox, Cow, Heifer, Steer, Sheep, or other Cattle'.[64] Some interpreted this broadly as outlawing any form of ill-treatment; in 1824 the Police Magistrate in Edinburgh fined a 'young man' one pound for 'setting a bulldog on [a] kangaroo' in Earl James's menagerie, persuaded that the 'the thing was wilfully and maliciously done'.[65] Other magistrates, however, interpreted the legislation more narrowly, insisting that bulls and other animals did not fall under the rubric 'cattle' and were not covered by the Act. In the Oxfordshire town of Thame, for example, where the tongue of a bull was 'literally torn in pieces by the dogs and exhibited by some cold-blooded wretch' on a plate, the magistrate, Mr Ashford, declined to prosecute, on the grounds that 'he did not consider bull-baiting to be a crime'.[66] His decision elicited a flurry of outraged correspondence from shocked locals expressing their horror that 'such a scene took place in the *streets of Thame* and passed unpunished'.[67]

The Warwick lion fight exposed some of these legal quandaries and highlighted the complexities and limitations of the 1822 Act. When the SPCA first heard that the contest was happening, Martin sent his agent Charles Wheeler to Warwick to try to prevent it. The Clerk of the Corporation and one of the two magistrates were, according to Wheeler, 'clearly of the opinion that the thing was illegal', and agreed to sign a warrant against it. The other magistrate, however, refused to support this decision, making it impossible to suppress the contest. Thwarted in the provinces, Wheeler returned to London in the interval between the first and second lion fights and tried to persuade the Lord Mayor to 'prevent Wombwell, the owner of the lion which [is] to be baited on Saturday, from having any place in Bartholomew Fair'. Wheeler believed that if Wombwell were denied his usual station at the fair, one of London's oldest and largest, he would cancel the proposed fight; the showman himself had reportedly said that 'if the consequence of the fight should be that he must be deprived of the annual harvest at Bartholomew Fair, he certainly should not allow the second experiment to be made'.[68] In the event, however, this second strategy also fell through, for the Lord Mayor claimed that he did not have authority over the letting of the land on which the fair took place and declined to interfere 'with the duties of other Magistrates'. The fight thus went ahead, and Wombwell exhibited at St Bartholomew's as normal, shamelessly using the lion bait to publicise his show. A disgusted William Hone described how the menagerie's façade featured prominently the wording 'NERO AND WALLACE; THE SAME LIONS THAT FOUGHT AT WARWICK', along with a picture of the second fight, in which 'a lion stood up with a dog in his mouth crunched between his grinders', blood running 'from his jaws'.[69] In this case, therefore, the law appeared powerless to prevent a public act of animal cruelty.

Conclusion

The Warwick lion fights were controversial and highly publicised events. Though unusual in themselves, they illuminated wider concerns over physical violence towards animals, which were starting to surface in the early nineteenth century. Occurring at a time when the parallel sports of bull-baiting and dog-fighting were receiving growing attention in the press, the courts and the Commons, the fights highlighted, in the most graphic manner, the abuses perpetrated against animals in the name of entertainment, raising questions about the moral responsibility of humans towards animals and the dangers of exposing spectators to scenes of brutality. The spectacles functioned as a barometer for public opinion on animal welfare in the 1820s, drawing condemnation from some MPs, the radical press, and nonconformist Christians (like the Quaker Samuel Hoare) while receiving active support from a relatively broad spectrum of society, including many members of the Warwickshire gentry. They also secured a lasting place in national memory, eliciting comment decades after they occurred; as late as 1873, a subscriber to RSPCA's monthly magazine, *The Animal World*, claimed acquaintance with a 'white bull dog' that had been 'one of the six' sent in to fight Nero, describing how the animal had been 'thrown out of the ring as dead, but picked up and purchased by Mr★★★'.[70]

As well as revealing the primary concerns of animal rights campaigners, the lion fights highlighted the constraints and complexities of existing legislation on animal protection, which, as Wheeler discovered to his chagrin, did not appear to cover lions. The 1835 Animal Cruelty Act in part remedied this situation, formally outlawing the baiting of any animal. The legal status of menagerie inmates, however, remained a source of contention throughout the nineteenth century, with several offenders escaping the law on the grounds that their victims did not qualify as 'domestic' animals and were not, therefore, covered by the existing statute unless explicitly engaged in 'baiting' with dogs. In 1874, for instance, when the RSPCA prosecuted keeper Frederick Hewitt for making hyenas jump through a burning hoop, presiding magistrate Mr Bruce reluctantly dismissed the case, on the grounds that 'no man, in the eye of the law, would ever think of classing a hyena kept in a menagerie a domestic animal'.[71] The RSPCA would work hard to close this loophole, particularly in the 1870s, bringing lions, hyenas and other species more firmly within the remit of the law.[72] The Warwick affair offers an early example of their objectives and tactics and foreshadows the way in which the courts, Parliament, and the press would shape the debate on animal cruelty in ensuing years.

The class dynamics of the Warwick lion fight also merit attention. The 1820s is usually seen as a decade in which the traditional sports of the lower classes were coming under attack as relics of an earlier and more brutal age.

The Warwick lion fight, however, does not quite fit this model. First, the vast majority of spectators were rural gentry, not labourers. Second, lion-baiting was not a traditional part of British popular culture, but an unabashedly commercial spectacle. There was, it is true, a tradition of exotic animal fights at early modern courts, including the court of James I in Britain, but the practice never evolved into a popular custom.[73] The baits were instead products of the newly commercialised leisure culture charted by Hugh Cunningham in *Leisure in the Industrial Revolution* (1980), motivated by profit, rather than longstanding local traditions, and accompanied by the same extensive and increasingly professional betting culture that permeated horse racing, cock-fighting, pugilism, and ratting.[74] This monetised, profit-making type of sport was different from the community bull-bait and very much a product of contemporary urban consumer culture. While the SPCA and its supporters directed much of their fire at the traditional pastimes of the poor, they were also forced to confront an array of new, more commercialised forms of cruelty that were products of, rather than anathema to, the modern age. The Warwick lion fight was a prime example of these changes.

Finally, the Warwick lion fights offer an intriguing insight into an unusual set of interspecies interactions. Though subjected to violence, both the dogs and the lions were treated quite explicitly as individuals, even canine or leonine celebrities, enjoying a close relationship with their human owners. The dogs were all endowed with names – some heroic, others less so. They were praised for their courage when they triumphed, 'waited upon by their keepers' after they left the arena, and lambasted in the press when they showed fear.[75] Turk, for example, was acclaimed for exhibiting 'gameness and strength almost beyond belief', while Tiger was condemned as 'a complete cur' after he refused to engage Nero.[76] As for the lions, they were heavily anthropomorphised in contemporary accounts and had their genealogy closely scrutinised. *Jackson's Oxford Journal* reported that 'Wallace was whelped at Edinburgh about six years ago' and 'was suckled and reared by a bull bitch' – a common practice in contemporary menageries. Nero, on the other hand, was the offspring of two lions 'caught together … on the coast of Barbary … being the first specimens of the black-maned species that were introduced into this country'.[77] Such details suggest that these were, on the one hand, privileged creatures endowed interesting pedigrees and quasi-human personalities. At the same time, however, both the dogs and the lions were valued commodities whose primary importance lay in their capacity to make money for their human owners. The female dog Nettle was 'sold … after the fight with Nero for 25 guineas' having acquitted herself well in the combat, while Wombwell, as we have seen, capitalised on Nero's and Wallace's subsequent notoriety to publicise his show, displaying a 'panoramic sketch of the scene of Wallace and the dogs' at St Bartholomew's Fair and exhibiting Nero 'in the very den in which he fought' when the menagerie visited Norwich.[78] The animals involved in the Warwick lion fight thus occupied an ambiguous position between respected companions and expendable sources of profit, giving rise to complex cross-species relationships.

Notes

1 *Liverpool Mercury*, 18 April 1825.
2 *Morning Chronicle*, 28 July 1825.
3 Lewis Gompertz, *Objects and Address of the Society for the Prevention of Cruelty to Animals* (London, 1829), 9.
4 Harriet Ritvo, *The Animal Estate: The English and Other Creatures in the Victorian Age* (Cambridge: Harvard University Press, 1987), 125–66.
5 Keith Thomas, *Man and the Natural World: Changing Attitudes in England 1500–1800* (London: Penguin, 1983), 180–2.
6 Jeremy Bentham, *An Introduction to the Principles of Morals and Legislation* (Oxford: Clarendon, 1907), 311.
7 Emma Griffin contends, for instance, that, despite the number of tracts written against animal cruelty in the eighteenth and early nineteenth centuries, few of these were actually read and fewer still properly digested. 'Sermons and tracts did not circulate in large numbers and their popularity (or lack of it) with the reading public is surely indicated by the fact that few were ever reprinted'. See Emma Griffin, *Blood Sport: Hunting in Britain Since 1066* (New Haven: Yale University Press, 2007), 143.
8 James Turner, *Reckoning with the Beast: Animals, Pain and Humanity in the Victorian Mind* (Baltimore: John Hopkins University Press, 1980), 15–38. See also Robert Malcomson, *Popular Recreations in English Society 1700–1850* (Cambridge: Cambridge University Press, 1973), 89–117. Malcolmson argues that 'industry' came to be increasingly prized in the late eighteenth century. In this new moral climate, popular recreations of all kinds came to be seen as unnecessary distractions from labour and an impediment to artisanal or industrial productivity.
9 See Hilda Kean, *Animal Rights: Political and Social Change in Britain Since 1800* (London: Reaktion, 1998), 30–31 and Diana Donald, '"Beastly Sights": The Treatment of Animals as a Moral Theme in Representations of London, c.1820–1850', *Art History*, 22 (1999), 516.
10 William Hogarth, *The Four Stages of Cruelty* (London, 1751).
11 *Bury and Norwich Post*, 2 June 1802.
12 *Morning Post*, 25 August 1825. For a more detailed history of the RSPCA, see Brian Harrison, 'Animals and the State in Nineteenth-Century England', *English Historical Review*, 88 (1973), 786–820. There were important overlaps between the SPCA and the contemporary abolitionist movement, both in terms of personnel and in terms of ideology, with both organisations emphasising the capacity to suffer as the primary criterion for moral consideration.
13 Gompertz, *Objects and Address*, 5.
14 *Berrow's Worcester Journal*, 31 March 1825.
15 *Bell's Life in London*, cited in *Morning Chronicle*, 4 April 1825.
16 *Hereford Independent*, cited in *Morning Post*, 28 June 1825. There were also a few spurious reports as other showmen exploited the publicity opportunity. An advertisement in the *Leeds Mercury*, for instance, claimed that George, 'the NOBLE LION' in Ballard's menagerie, would 'shortly after Leeds Fair, combat Six Bull Dogs, of which due Notice will be given'. See *Leeds Mercury*, 25 June 1825.
17 *Morning Chronicle*, 28 July 1825.
18 *Jackson's Oxford Journal*, 30 July 1825; *The Examiner*, 24 July 1825.
19 *Caledonian Mercury*, 1 August 1825.
20 Ibid.
21 *Jackson's Oxford Journal*, 30 July 1825.
22 'Great Fight between the Lion Nero and Six Dogs', Warwick County Records Office, Warwick (hereafter WCRO), CR1097/330/p. 353 part VII.
23 *Morning Chronicle*, 28 July 1825.
24 Ibid.
25 *Jackson's Oxford Journal*, 30 July 1825; *Caledonian Mercury*, 1 August 1825.

26 *Morning Chronicle*, 28 July 1825.

27 *Jackson's Oxford Journal*, 30 July 1825.

28 *Caledonian Mercury*, 1 August 1825.

29 Copy of a handbill advertising the fight, reproduced in 'Great Fight between the Lion Nero and Six Dogs', WCRO, CR1097/330/p. 353 part VII.

30 *Jackson's Oxford Journal*, 30 July 1825.

31 *Examiner*, 31 July 1825.

32 *Jackson's Oxford Journal*, 30 July 1825.

33 *Liverpool Mercury*, 29 July 1825.

34 *Jackson's Oxford Journal*, 30 July 1825.

35 *Caledonian Mercury*, 2 November 1822.

36 *Jackson's Oxford Journal*, 6 August 1825.

37 *York Herald*, 29 Sepember 1827.

38 *Jackson's Oxford Journal*, 6 August 1825.

39 *Morning Post*, 1 August 1825.

40 *Jackson's Oxford Journal*, 30 July 1825.

41 William Hone, *The Every-Day Book* (London, 1826), 994–95 (26 July 1825).

42 *Caledonian Mercury*, 1 August 1825; *Examiner*, 31 July 1825.

43 *The Standard*, 11 July 1828. Emma Griffin notes the increasingly emotive depictions of baited bulls in the early nineteenth-century and suggests that such portrayals were exaggerated for propagandistic purposes. See Griffin, *Blood Sport*, 144.

44 *Liverpool Mercury*, 29 July 1825.

45 *Liverpool Mercury*, 8 April 1825.

46 *Liverpool Mercury*, 29 July 1825.

47 *Jackson's Oxford Journal*, 5 November 1825.

48 *The Times*, 28 July 1825.

49 On the many different ways of classifying and categorising animals in nineteenth-century Britain, see Harriet Ritvo, *The Platypus and the Mermaid and Other Figments of the Classifying Imagination* (Cambridge: Harvard University Press, 1997). Ritvo argues that pets enjoyed a privileged relationship with humans, valued for their affective attributes rather than their material uses (193–94). As a wild animal reared in captivity and exhibited for profit, Nero straddled the boundary between domestic and wild.

50 'Wild Beast Statistics', *North Wales Chronicle*, 29 July 1834. Wombwell divulged this information in an interview with the *Dumfries Courier*.

51 *Blackwood's Edinburgh Magazine*, 472 (February 1855), 189–90. For further discussion of human-animal interaction in travelling menageries, see Helen Cowie, *Exhibiting Animals in Nineteenth-Century Britain: Empathy, Education, Entertainment* (Basingstoke: Palgrave Macmillan, 2014), 52–76.

52 *Liverpool Mercury*, 29 July 1825.

53 Hone, *Every-Day Book*, 998 (26 July 1825).

54 Parl. Debs. (series 2) vol. 12, cols. 1002–13 (11 March 1825).

55 'Revolting Sanguinary Sport', *The Animal World: An Advocate of Humanity*, 7 (January 1876), 2.

56 William Hamilton Drummond, *The Rights of Animals: And Man's Obligation to Treat Them with Humanity* (Dublin, 1838), 104.

57 Parl. Debs. (series 2) vol. 12, cols 1002–13 (11 March 1825).

58 Parl. Debs. (series 2) vol. 14, cols. 647–52 (21 February 1826).

59 Parl. Debs. (series 2) vol. 12, cols. 1002–13 (11 March 1825).

60 *Bell's Life in London*, 3 July 1825. Youatt thought that training dogs to attack and bite encouraged rabies, the causes of which were still in dispute in the 1820s. See Neil Pemberton and Michael Worboys, *Rabies in Britain, 1830–2000* (Basingstoke: Palgrave Macmillan, 2007), 19–26.

61 Parl. Debs. (series 2) vol. 12, cols. 1002–13 (11 March 1825).

62 Parl. Debs. (series 2) vol. 14, cols. 647–52 (21 February 1826).

63 'A bill to consolidate and amend the several laws relating to the cruel and improper treatment of animals', Parl. Papers (Commons), 1835 (93), 5 (emphasis mine). Enacted on 9 September 1835 as 5 & 6 Will. IV, c. 59.

64 *An Act to prevent the cruel and improper treatment of cattle*, 3 Geo. IV [1822], c. 71.

65 *Caledonian Mercury*, 8 April 1824. It is possible in this instance, however, that the prosecution related to damage to property rather than cruelty to animals.

66 *Jackson's Oxford Journal*, 5 November 1825; *Bell's Life in London*, 13 March 1826.

67 *Jackson's Oxford Journal*, 3 December 1825. Emphasis in original.

68 *Morning Chronicle*, 30 July 1825. The working-class paper, *Bell's Life in London*, opposed this penalty, which it viewed as a form of class prejudice. As the paper expressed it: 'It is said that Mr Wombwell is to be denied standing room for his menagerie, in consequence of the late lion fight at Warwick. Upon the same principle, a fox hunter who has ridden his horse to death should be excluded from civilised society'. See *Bell's Life in London*, 14 Aug. 1825.

69 Hone, *Every-Day Book*, 1198 (26 July 1825).

70 'One of the Six', *The Animal World*, 4 (January 1873), 14.

71 *Daily News*, 9 December 1874.

72 In 1870, for instance, the RSPCA successfully prosecuted two German men for hitting 'a blind and emaciated bear' with a stick; in 1882 it gained a conviction for 'torturing' a bear with a 'ring through [the] lip'; and in 1887 it secured convictions for 'beating, kicking, stabbing' a monkey and a dromedary. See *The Animal World*, 1 (December 1870), 40; 14 (August 1883), 115; 19 (August 1888), 116. Not all prosecutions succeeded however, and the status of menagerie inmates remained uncertain. In 1894, for instance, a judge at Westminster Police Court dismissed a case brought by the RSPCA against the keepers at the Westminster Aquarium for 'cruelly beat[ing], ill-treat[ing], abus[ing] and tortur[ing]' five caged lions, arguing that 'these lions [were] not domestic animals', and were not covered by the Animal Welfare Act. See 'Lion Beating at the Westminster Aquarium', *The Animal World*, 25 (August 1894), 116.

73 Eric Baratay and Elisabeth Hardouin-Fugier, *Zoo: A History of Zoological Gardens in the West*, tr. Oliver Welsh (London: Reaktion, 2002), 17–28.

74 Hugh Cunningham, *Leisure in the Industrial Revolution c.1780–1880* (New York: St Martin's Press, 1980), 9. On the sport of ratting, see Neil Pemberton 'The Rat-Catcher's Prank: Interspecies Cunningness and Scavenging in Henry Mayhew's London', *Journal of Victorian Culture*, 19 (2014), 520–35.

75 *Caledonian Mercury*, 1 August 1825.

76 *Jackson's Oxford Journal*, 30 July 1825, 6 August 1825. The famous ratting dog Billy even had his portrait painted in 1827, and received an 'elegant collar' from his supporters in 1822, upon which was inscribed 'his name and wonderful exploits'. See *Liverpool Mercury*, 28 December 1827; *Morning Post*, 11 October 1823.

77 *Jackson's Oxford Journal*, 6 August 1825.

78 Ibid.; *Bell's Life in London*, 4 September 1825; *Bury and Norwich Post*, 21 October 1825.

Bibliography

Primary sources

Archival Sources

Warwick County Records Office, Warwick
CR1097/330/p. 353 part VII 'Great Fight between The Lion Nero and Six Dogs'

Newspapers and Periodicals

The Animal World: An Advocate of Humanity (London: S.W. Partridge, 1869–1970)

Bell's Life in London

Berrow's Worcester Journal

Blackwood's Magazine

The Bury and Norwich Post

The Caledonian Mercury

Daily News

The Examiner

The Hull Packet

Jackson's Oxford Journal

The Leeds Mercury

The Liverpool Mercury

The Morning Chronicle

The Morning Post

North Wales Chronicle

The Standard

The Times

The York Herald

Official Publications

3 Geo. IV [1822], c. 71: *An Act to prevent the cruel and improper treatment of cattle.*

Great Britain. Parliamentary Papers (Commons), 1835 (93): 'A bill to consolidate and amend the several laws relating to the cruel and improper treatment of animals'.

Great Britain. Parliamentary Debates (series 2), vols 12 (1825), 14 (1826).

Bentham, Jeremy, *An Introduction to the Principles of Morals and Legislation* (Oxford: Clarendon, 1907).

Gompertz, Lewis, *Objects and Address of the Society for the Prevention of Cruelty to Animals* (London, 1829).

Hamilton Drummond, William, *The Rights of Animals: And Man's Obligation to Treat Them with Humanity* (Dublin, 1838).

Other Printed Sources

Hogarth, William, *The Four Stages of Cruelty* (London, 1751).

Hone, William, *The Every-Day Book* (London, 1826).

Secondary sources

Baratay, Eric, and Elisabeth Hardouin-Fugier, *Zoo: A History of Zoological Gardens in the West* (London: Reaktion, 2002).

Cowie, Helen, *Exhibiting Animals in Nineteenth-Century Britain: Empathy, Education, Entertainment* (Basingstoke: Palgrave Macmillan, 2014).

Cunningham, Hugh, *Leisure in the Industrial Revolution c.1780–1880* (New York: St Martin's Press, 1980).

Donald, Diana, '"Beastly Sights": The Treatment of Animals as a Moral Theme in Representations of London, c.1820–1850', *Art History*, 22 (1999), 514–44.

Griffin, Emma, *Blood Sport: Hunting in Britain Since 1066* (New Haven: Yale University Press, 2007).

Harrison, Brian, 'Animals and the State in Nineteenth-Century England', *English Historical Review*, 88 (1973), 786–820.

Kean, Hilda, *Animal Rights: Political and Social Change in Britain Since 1800* (London: Reaktion, 1998).

Malcomson, Robert, *Popular Recreations in English Society 1700–1850* (Cambridge: Cambridge University Press, 1973).

Pemberton, Neil, 'The Rat-Catcher's Prank: Interspecies Cunningness and Scavenging in Henry Mayhew's London', *Journal of Victorian Culture*, 19 (2014), 520–535.

Pemberton, Neil, and Michael Worboys, *Rabies in Britain, 1830–2000* (Basingstoke: Palgrave Macmillan, 2007).

Ritvo, Harriet, *The Animal Estate: The English and Other Creatures in the Victorian Age* (Cambridge: Harvard University Press, 1987).

Ritvo, Harriet, *The Platypus and the Mermaid and Other Figments of the Classifying Imagination* (Cambridge: Harvard University Press, 1997).

Thomas, Keith, *Man and the Natural World: Changing Attitudes in England 1500–1800* (London: Penguin, 1983).

Turner, James, *Reckoning with the Beast: Animals, Pain and Humanity in the Victorian Mind* (Baltimore: Johns Hopkins University Press, 1980).

Part III

Self and other

Identification and classification

8 Inveterate travellers and travelling invertebrates

Human and animal in Enlightenment entomology

Dominik Hünniger[1]

On 19 August 1768, the English naturalist John Ellis wrote enthusiastically to Carl Linnaeus about the splendid equipment aboard the HMS *Endeavour* on James Cook's first circumnavigation.[2] Interestingly, the first species Ellis mentioned in his letter were invertebrates, and the tools for their capture: 'they have all sorts of machines for catching and preserving insects'.[3] This letter is amongst the most quoted in the literature on Joseph Banks and the history of British explorations during the long eighteenth century.[4] Some of its details have so far escaped historical scrutiny, particularly those concerning animals and the relations between humans and other creatures.

This chapter is inspired by Erica Fudge's work on the history of the relationship of humans with other animals. Both animals and humans are featured here as 'world producing beings',[5] and both had their specific roles in the making of knowledge. It remains evident though that the character of action and the means of knowledge production are very different in this interaction. All species – humans and invertebrates – were travelling widely but not for identical reasons. Insects did this very rarely, if at all, of their own volition. They were caught and preserved but sometimes also travelled accidentally as stowaway passengers. Nevertheless their specific behaviour and the difficulties in catching, preserving and ordering the insect world inspired the humans in this story to develop a set of tools and methods in their interaction with insects. In this process, these insects could hence be seen as 'subjects to be negotiated with' to use Erica Fudge's phrase.

By looking at entomology and its practitioners as well as the objects of their study in the last decades of the eighteenth century, this chapter expands current research on interspecies interaction. Its analysis of contemporary travel literature in conjunction with systematic natural history publications offers new insights into the representation and materiality of different species – humans and insects – in their interaction through one and the same historical process: the establishment of a novel field of scholarly interest, entomology.

While botanical exchange and cultural encounters in the context of European exploration and colonialism have been studied quite extensively over the past decades, interspecies encounters and fauna – especially 'non-charismatic micro-fauna'[6] – are largely invisible in the recent literature.[7] Yet, the growing interest in human-animal relations and environmental history is producing new perspectives on the history of exploration and collecting. Likewise, historians of natural history collections and taxonomy also stress the importance of looking into historic zoological collections for current taxonomic and biodiversity research in the life sciences.[8]

Unfortunately the history of entomology as an academic discipline has not yet received due attention, especially for the formative years around 1800.[9] The growing interest of the humanities in humans' interaction with other animals has, however, sparked some interest in insects in European history.[10] This chapter will expand this research by looking into the inter-European exchange not only conducted in written form but also manifested in actual collaboration at the well-known centres of collecting (Paris and London) as well as at many smaller natural history cabinets existing throughout Europe. This also included the exchange of duplicates and even whole collections. The collected species themselves became a commodity, and the epistemic status of the animals for systematics was very much debated. The argument evolved around the increasing conviction that one had to inspect specimens personally in order to sufficiently describe and include them within one's systematics. The use of second-hand descriptions or visual depiction was increasingly rejected as un-systematic by many naturalists.[11]

Two places in particular served as spaces of encounter: the natural world where humans collected insects and the collections where the insects were pinned, displayed and occasionally dissected. Both spaces shaped the ways in which humans could interact with each other and with insects in specific forms. They also made humans reflect on the corporeal as well as epistemic form and status of specimens. Thus, they provide ideal spaces for a study on human-animal encounters.

Insects and humans in the field

As the demand for new specimens from home and abroad grew, the literature on how to obtain them expanded as well. A large number of manuals and handbooks on how to catch, preserve and order invertebrates were published in many European languages during the second half of the eighteenth century. The (in)famous naturalist and explorer Johann Reinhold Forster gave a detailed account on how to catch insects in his *Catalogue of the Animals of North America* (1771). This publication emerged from his work on a collection by the Hudson's Bay Company sent to London in 1771.[12]

It is, however, apparent that Forster's own experiences as a collector must have informed this account, as it is extremely detailed and very technical. His descriptions of the instruments to catch insects are essentially blueprints for manufacturing these 'little tools of knowledge'.[13] He described forceps or scissor nets to catch the live animals, and pincushions, different sized pins, and store boxes 'to put therein the insects caught in the various excursions'.[14] Forster also mentioned a large mosquito net that could be obtained ready-made in London. Furthermore, he described different pinning techniques for different genera of invertebrates.[15]

Forster also admitted that the animals of course did not offer themselves easily to the instruments of the collector: butterflies proved especially problematic as they beat their wings against the forceps and thereby rubbed off some of their wings' scales. Clearly, the animals' behaviour put their collection as scientific objects in jeopardy. Without the scales, there was no colour pattern, and hence one characteristic of classification and differentiation was lost. Forster therefore found it 'necessary to give these creatures, when in the forceps or net, a gentle squeeze at the insertion of the wings in the body'.[16] Forster however recommended not only force but also gentleness in dealing with insects: after the excursion, the creatures should be put 'on a large pincushion, by which means they will be enabled to rest their feet on and this will prevent their fluttering'. The rather technical and matter-of-fact quality of the manual makes it very likely that this procedure was solely in the interest of the collector who wanted to preserve the whole specimen and its characteristics.[17]

Forster was slightly less concerned about keeping other genera alive: 'Beetles, and many of the half-winged insects, may be dipped in the preparing liquor, which will kill and put them soon out of pain and prevent small insects from destroying them'.[18] Forster had no problem in putting beetles, crabs, millipedes, spiders, worms and scorpions into a bottle of 'rum or rack' in order to send them overseas. This passage is again ambivalent: animal pain was to be avoided, but the main reason for all these measures was the preservation not of the live animal but of the specimen in the collection. This was indeed the only thing that mattered for the systematic naturalist, as Sheila Wille has also recently argued. Indeed, it was also peculiar to Forster to recommend meticulous recording of the time and place when and where specimens were found and caught. Forster recommended paying attention to 'the plant or food it lives upon, its changes, and what animals feast again upon the insect, and other such particularities'.[19]

In contrast to Forster's demand for a detailed record of the animals' habitat and life cycle, the contemporary entomological taxonomy handbooks, like Johann Christian Fabricius' *Systema Entomologiae* or Guillaume Antoine Olivier's multivolume *Entomologie*, barely mentioned these specifics.[20] The reason for this neglect was not only that many of the collectors on whose specimens these systematics were based had not recorded such details. Even more importantly, late eighteenth-century systematic entomology was concerned first and foremost with taxonomy based on morphological characteristics rather than habitat. Yet, a few entomologists criticised this narrow approach to a seemingly much more complex phenomenon: 'Although the systematic arrangement of insects has of late been prodigiously advanced, the philosophical study of their economy does not appear to have been equally cultivated'.[21] The founder of the Linnean Society, James Edward Smith, also stressed that many of the recent taxonomies had only paid attention to the fully developed insect and not to metamorphoses as they had only studied insects in the cabinets. Smith's plea for more detailed attention to the whole life cycle of insects is mirrored by a growing attention to questions of transformation, life cycles, and economic and medical usage that were increasingly found in another medium: the (specialised) scholarly journal in (at least) German-speaking Europe.

Concerns about the preservation of animal bodies also featured prominently in discussions about the exchange of specimens among natural historians. At the end of the eighteenth century the number of species that were traded and circulated rose dramatically. This was not only an outcome of the growing amount of material coming in from European colonies and newly explored areas of the globe, but also of exchanges between metropolis and periphery. In a letter to Banks, the German naturalist and St. Petersburg professor Peter Simon Pallas sent 'a Box of Insects most peculiar to our Country' and hoped they would 'prove for a good part agreeable by their novelty to your discerning eye'.[22] Pallas asked for similar favours in return, especially insects. Deploying Linnaean terminology, Pallas was desperate to receive specimens from distant shores: 'Coleoptera & Hymenoptera are most in favour with me; but any other exotic Insects will be most gratefully acknowledged'.[23] Pallas, like any other systematic naturalist, relied heavily on the plentiful arrival of new specimens in Europe whether they had been collected by fellow naturalists, sailors, merchants or indigenous people.[24] In order to obtain them, Pallas had either to travel himself – which he did – or as can be seen from the letter just quoted to enter into the community of scholars and its gift economy.[25]

As many naturalists were wholly dependent on these exchanges, they reflected not only on the preservation and storage of the insects in their collections but also on how to make them fit for travel. Numerous manuals on how best to collect, store and send animals were published in many different languages. The German scholar, Johann Beckmann – also a pupil of Linnaeus – reflected on the welfare of the insects during these procedures in his manual entitled 'A more convenient arrangement of insect collections'. Beckmann's paper was published in one of the most important natural history journals in German-speaking Europe: the *Transactions of the Berlin Society of Friends of Nature Research* (*Beschäftigungen der Berlinischen Gesellschaft Naturforschender Freunde*).[26] The article is a great example of the creativity and zeal with which natural historians tried to improve their science. Additionally, Beckmann focuses especially on techniques and material. This comes as no surprise, as he is considered the 'inventor' of *Technologie* as an academic discipline in Germany.[27]

Beckmann had been frustrated by the weak layout and material of regular insect drawers because they proved inadequate to protect specimens. Therefore Beckmann suggested a new model that he himself had invented and tested. In best enlightened, improving style, Beckmann laid out the problems and difficulties of current practices and materials. His main concern was the steadfastness of the pin, and he mentioned that many of his colleagues were heating the pin in order to kill the insect faster. Beckmann rejected this method, first, because many insects could not thereby be killed and, second, because the pin was prone to break when heated. Interestingly, Beckmann had also experimented with heat but was very reluctant to recommend it as he found it cruel and only administered it with distaste.[28] This is one of the rare instances when contemporary entomologists expressed concern with their 'objects' of study: discussions of entomologists' emotions when handling insects occurred very rarely.

The rest of Beckmann's paper consists of a meticulous description of how to build his ideal drawer. Like Forster's manual, Beckmann's paper could indeed be used as a blueprint to manufacture these instruments and gadgets. Beckmann's main concern was how to keep other insects, especially collection 'pests', out of the cabinet. His suggestion was to keep the specimens in different cases, placing them in the special drawer only after it was certain they were unharmed. Subjecting specimens to the heat of the sun in summer – or using heating in winter – was also recommended for saving the specimens from other insects that would feed on them. Beckmann was happy to report that he was able to preserve specimens in his collection for 'eight, ten, twelve years' with this method. Here another form of interspecies encounter in the collections becomes apparent: the imminent danger of destruction at the hands of other insects. As the animals in the collections were no longer alive, damage done by the living was an ever-present danger and occurred often. It could only be prevented with great difficulty and much effort.[29]

Entomologists had to fight off some species of insects in order to keep the dead bodies of others intact in their collections. Beckmann's drawer manual was one attempt at tackling this problem. Additionally, his invention could also be used by the naturalist for posting or travelling with the collection or parts of it. The last part of Beckmann's paper includes a detailed description of how to build a device to protect the drawer from the perils of movement when it was used in the gift economy of species exchange and on the road while the entomologist was collecting. Accordingly, Beckmann ended his paper affirming that he 'safely transported some hundred insects for some hundred miles in this manner aboard the rugged German post stage coaches'.[30]

The cabinets of natural history

In their seminal collections of essays on the 'Sciences in Enlightened Europe', William Clark, Jan Golinski and Simon Schaffer have described Paris and London 'as the Enlightenment's material centers'.[31] Indeed, the rich natural history collections in the British and French metropolises were instrumental to the production and dissemination of knowledge and the making of the scientific disciplines around 1800. They were 'centres of calculation', where dispersed information and materials were accumulated and compared in order to generate knowledge of the (natural) world that was eventually used to dominate it.[32]

It has long been argued that collections and first-hand encounters with specimens and objects in particular were among the most important features of eighteenth-century natural history. The reliability of travel reports was a matter of continuous debate in the eighteenth century.[33] In zoology and botany, at least, first-hand observation was favoured over second-hand encounters. This becomes apparent in the praise of British natural history collections by the Danish entomologist Johann Christian Fabricius. When he visited London in 1782, he described their great importance for European naturalists because the 'abundance of unknown and strange natural specimens [... was] necessary to give my knowledge of natural history in general and my insect system in particular the proper scope, as all real acquaintance with nature is based on knowledge and the comparison of species and every system must be supported by this'.[34]

The collections were characterised as the space for interspecies encounter necessary for the development of the discipline. Fabricius' strong emphasis on the importance of the specimens themselves for systematic natural history is particularly telling and echoes the great attention paid to collecting and preserving already discussed above.

Fabricius was not alone in realising that the British metropolis was the central contact zone for continental European natural historians in the eighteenth century. German, Scandinavian, Dutch and French travellers and naturalists all praised and made use of the easy availability and rich diversity of the material collected by British explorers and naturalists at home and especially abroad. Furthermore, British collectors welcomed foreign natural historians without any need for formal introduction, as Fabricius' friend Guillaume Antoine Olivier acknowledged in the preface to his entomological handbook.[35] Accordingly, John Gascoigne called Joseph Banks's house his 'home-cum-research-institute at Soho Square'.[36] It was practically a privately funded institution with its own employees and guest researchers. These guests, however, not only took advantage of the facilities but contributed expertise, knowledge and contacts.

Furthermore, British explorers and collectors relied on expertise from abroad and advertised their rich collections not only in the travel accounts but also in their correspondence. Thus wrote Joseph Banks to the French naturalist Louis-Léon-Félicité de Brancas, duc de Lauraguais: 'The Number of Natural productions discover'd in this Voyage is incredible: [...] 500 fish, as many Birds, and insects Sea and Land innumerable'.[37] The volume of insects brought back to Europe seemed to be so large that it was not even possible to give an exact number for them. Consequently, European naturalists themselves flocked in great numbers to the Empire's metropolis to see, describe and systematise what came back with the explorers' ships. Indeed, the exchange of species as well as of other goods is the subject of many a letter from European naturalists to Joseph Banks.[38]

At the same time, from the 1770s onwards, specialisation in specific parts of the natural kingdoms seems to have grown, and many naturalists tried to establish very specific systematics. These, of course, could only be created when specimens were available. This effort proved difficult for many scholars on the Continent as they could not complete their systematic work at home without direct access to natural materials. The rapidly spread news that circumnavigators had returned to London carrying specimens had an electrifying effect on Continental scholars. Fabricius described this in a letter to Banks: 'I am very happy to hear of the many fine things Mr Forster has brought home and of the many plants you have gott from the Cape, wishing only to be once more with you to look and stare. Has he brought any Insects? I want much to see them'.[39]

Fabricius' letter highlights the longing for novelty and the pleasure he seemed to have enjoyed with Banks's collection. However, the study of insects was still ridiculed and many entomologists were concerned by the lay perception of most insects as detestable and vile. In true Enlightenment fashion, almost all authors readily discarded this notion and focused instead on the utilitarian aspects of insects. Dru Drury, for instance, identified a lack of knowledge and familiarity as the main reasons for people's misconceptions of invertebrates: 'We are too prone to think every thing noxious and unnecessary, if we are not fully acquainted with its use'.[40] Drury specified this claim with examples and also encouraged his readers to study their local fauna more closely in order to find out more about possible uses and thereby also replace 'foreign' produce with local. Here, Drury echoes a well-known Enlightenment sentiment, so brilliantly analysed by Lisbet Koerner in her case study on Linnaeus' attempts to cultivate neophyte plants in Sweden.[41] Utility and the exploitation of local resources was the mainstay of mercantilist and cameralist theory, and references to this were apparently thought to be successful in convincing critics of the importance of entomological studies.

This could however be held against the global aspirations of eighteenth-century entomology. Samuel Johnson specifically used the example of insects to question the benefits of the *Endeavour* voyage. When James Boswell tried to convince his friend that the number of new species collected on the voyages alone was a true sign of success, Johnson allegedly merely replied: 'Why Sir, as to insects, Ray reckons of British insects twenty thousand species. They might have staid at home and discovered enough in that way'.[42] In fact, the sheer quantity of possible species to encounter was one of the main anxieties but also motivations of the practitioners of eighteenth-century insect studies.

The very tangible fascination with the multitude of the miniscule was traditionally explained by aesthetic and educational as well as religious arguments. Drury uses terms such as 'wonderful', 'extraordinary' and 'astonishment' very frequently. Additionally, the aesthetic aspects of entomology and insects themselves were a function of the enlightened zeal for knowledge. Hence entomology could become simply another form of worship or acknowledging the power of a divine being: 'If their shape and beauty are capable of attracting our notice, their ways of living are no less adapted for exciting our admiration; and the more we enquire into their nature and history, the more occasion we shall find for confessing this great truth, nothing is created in vain'. The most important argument to vindicate insect studies in the eighteenth century was that insects 'proclaim the wisdom, goodness, and omnipotence of their great Creator'.[43]

The entomologists seemed to have had to defend their discipline until well into the nineteenth century. William Kirby and William Spence lamented this and envied the situation of botanists whose subject was treated as fashionable and useful because of its immediate connection to medicine. The practice of entomology however was 'so strongly associated with that of the diminutive size of its objects, that an entomologist is synonymous with every thing futile and childish'.[44] Contrary to Kirby and Spence's claims, insects and their bodies had been successfully employed in silk production, manufacturing of dyes, and medical practice.

The smallness of insects and the resulting difficulty of their study was certainly one of the most discussed issues in late eighteenth-century entomology. The years around 1800 witnessed fierce debates on taxonomy and systems revolving around the size and features of the insects.[45] This influenced the exact place of individual specimens in the cabinets as well as how they were treated by their human collectors and systematisers. The small sizes of the 'objects' of scrutiny compelled the human observer to make use of a wide variety of instruments and technologies.

One of the most important attempts to find a general system in late eighteenth-century entomology was Fabricius' 'cibarian system'.[46] Fabricius' method of classification concentrated on a very specific part of the insect. Like his teacher Linnaeus, Fabricius classified the insects hierarchically into species, genera and classes. However, he departed from his teacher in the characteristics that were the basis of this classification: instead of using the wing venation, as Linnaeus and many other earlier entomologists in Britain and on the Continent had done, he based his system on the mouthparts.[47]

In a paper first published in the *Journal of the Copenhagen Natural History Society* in 1790, Fabricius explained the importance of systematics and the tasks of the natural historian.[48] First, the soundness and certainty of the genera were considered the most important prerequisite for any system. This was the only way to determine the species. Hence, the main work of the natural historian was to discover and fix the characteristics of the genera. Fabricius argued that it was of the utmost importance to use only characters in taxonomy that are discernible in every species of the genus. For him the mouthparts were the most natural distinctive characteristic separating the insect genera, since their forms were determined by the way every insect feeds. This, in turn, influenced the insects' whole mode of living and *oeconomy*.[49] Consequently, for Fabricius, those natural species that display the same nourishment and way of life must necessarily belong to the same genus.

Fabricius was certainly aware of the difficulties this entailed: the mouthparts were tiny, hard to recognise without enlargement and 'in truth often hard to discern and to describe'.[50] Still, Fabricius maintained that the flowering parts in plants – the characteristics chosen by Linnaeus in his classification method – were even smaller. Only experience would eventually help the dedicated taxonomist to find the characters more easily.

Other entomologists were far less confident that the cibrarian system could be used for meaningful taxonomic encounters with insects. Adrian Haworth, who wrote one of the first English reviews of the entomological literature called it a 'misfortune' that Fabricius had based his system 'upon such minute parts'.[51] Haworth expressed concerns about the use of a microscope as it would need a 'careful dissection, by a skilful and accustomed hand'. Interestingly, Haworth continued to praise Fabricius' achievements in advancing entomology far beyond those of Linnaeus, despite his reservations about Fabricius' methods.

In addition to the travel necessary to acquire new specimens in the various European collections, it was necessary for Fabricius to have his reference system with him. Indeed, he was accompanied on his travels by his insect samples: 'I keep these parts [i.e. the mouthparts] in order to be able to compare'.[52] To allocate every species its place in the 'system', Fabricius needed to take animals apart and have samples of mouth parts with him when travelling. In their encounter with humans, insects were disassembled, relocated and brought into a system.

This disassembly was one of the most highly contested issues in entomology around 1800. In the *Magazin für Insektenkunde* – one of the first specialised journals solely concerned with insect studies – Fabricius and its editor, Johann Karl Wilhelm Illiger, engaged in a debate that revolved around systematics and the status of specific body parts of insects as a marker of differentiation. Illiger opened the debate with a review of Fabricius' 'cibrarian system'. Although he praised Fabricius' extraordinary achievements, he was very concerned because applying the Fabrician method would mean 'sacrific[ing] precious insects'.[53] Illiger maintained, as it was so difficult to discern the mouth parts, one had to separate the head from the thorax in order to observe the minute parts clearly. This process even had to be repeated sometimes and would lead to the destruction of many specimens. Illiger was very sceptical about the practicality of this method and questioned its accuracy entirely. His reasoning was based on contemporary developments in natural history that favoured experience and experiment: an accurate system had to be tested and practiced repeatedly.[54] Consequently, lack of experience was Illiger's main concern: only the most experienced practitioner would be able to discern the parts easily.

Additionally, Illiger argued for a broader approach to entomological systematics that also took other outer and inner body parts as characteristics for differentiation. Here, Illiger's position is somewhat contradictory as an analysis of inner body parts would also involve dissection. Illiger's paper in general oscillates between a strong admiration for the well-known entomologist and an attempt to establish a different system. Throughout Fabricius remains an authoritative figure for Illiger as he described so many species. Collections especially continued to be viewed as the most important space for insect systematics, and Illiger praised Fabricius' oeuvre for including references to all the collections in which he had seen specimens. The last paragraphs of Illiger's paper reveal an interesting potential motive behind his argument. His ideal prerequisite for a fully conceived insect systematics was a comprehensive and all-embracing insect collection that would consist of a complete representation of all species. He praised the collections of his patrons in Braunschweig as being very near completion and appealed to all 'friends of insect studies' to send in specimens to complete the collections.

Of course, Fabricius defended his system and had a chance to do so in the second volume of Illiger's periodical. As in earlier papers, he stressed the superior epistemic status of the mouthparts because they were the means of the genera's nourishment. For Fabricius this was the most important characteristic of an animal's life and thus the defining marker of differentiation between genera. In his reply to Illiger, Fabricius especially countered two main points of criticism: first, the difficulties of observation and systematisation and second, dissection and destruction. Contrary to Illiger's claim, Fabricius maintained he had seen many collections in which his system was adopted. He even asserted that he had seen many collections in which the mouthparts had been dissected and put on paper for each genus. Collections and their prototypical function were apparently as important for Fabricius as they were for Illiger. Fabricius also maintained that he only needed a simple magnifying glass in order to ascertain the characters and even 'some women have made great advances in these investigations and preparations'.[55] The unmistakably gendered rebuke to Illiger was certainly effective as this expressed a strong contempt towards Illiger's alleged inability to perform a task that 'even some women' could master.

Fabricius also addressed the issue of destruction. He admitted damage was necessary but considered this to be not a problem of the system but of collection: the very important collections would certainly have enough specimens. Again Fabricius tried to counter Illiger's argument in kind. He also disregarded any concern for individual specimens. An interpretation of this indifference as disrespect for individual animal wellbeing would however be rather problematic. As far as Fabricius was concerned, insects in the collections were specimens and served the purpose of establishing a comprehensive knowledge of the structure of the natural world. As questions of extinction and what we would now call biodiversity were just being raised, Fabricius was like most of his contemporaries convinced 'God's nature' was plentiful and endless.[56]

Illiger had also argued that not many entomologists had followed Fabricius' system. Of course Fabricius addressed this rather delicate point. Indeed, Fabricius' system had not found European-wide recognition but he could claim that many critics of his system, such as Kirby, did depict and use the mouthparts for certain genera in their insect studies.

Finally, Fabricius used another argument in which a distinction was made between two groups of entomologists. This argument is particularly interesting as it foreshadows a development that had only recently begun when

Fabricius was writing: the separation of lay and academic natural history and the development of the disciplines arising from earlier, broader study of the natural world. Fabricius denounced those 'who study insects not as a science but as a hobby'.[57] According to Fabricius they were dilettantes and refrained from a thorough pursuit of insect studies because of a prejudice that it was impossible to observe the mouthparts. As can be seen here, humans not only encountered other animals in the collections, but themselves split into at least two different 'species'. During the process of developing entomology as a science, collections and their specimens formed the arena from which the lay enthusiast and the scientist emerged.

Conclusion

How insects and humans were affected by their interspecies encounter was – literally – vitally different. Nonetheless, a common history of both species needs to take into account that humans conceptualise and experience their encounter with other species in myriad ways. Dominion or compassion – the two guiding principles between which (human) animal histories seem to oscillate – are just two possibilities.

Late eighteenth-century entomology required specimens, instruments and the mobility of invertebrates and naturalists. The resulting encounter produced new forms of knowledge that came at a price in more than one sense. Exploration, travel and work in the collection of course required constant funding. The insect as object of study underwent different stages in its encounters with human beings. First, the primary collector took it from its natural habitat. Then, the person in whose collection it ended up mounted it and put it into some form of a box, drawer or cabinet. Finally, the natural historian may have disassembled the specimen and only used one small part of the insect's body in his or her further studies and taxonomy. These roles might be filled by different people or the same person at different times.

Through this process, the humans in the story indeed produced a new world of entomological systematics inspired by Linnean taxonomy, fuelled by a constant supply of insect specimens travelling around the globe. The material context of this encounter relied heavily on 'tools of knowledge' but also on the distribution of labour as well as on interspecies encounter. Entomology as a science emerged as a multifaceted process with a wide array of practitioners who themselves struggled for recognition. The ensuing debates almost always revolved around the objects of study and their bodies.

Notes

1   This paper and its title are inspired by Richard I. Vane-Wright, 'Johann Christian Fabricius. Classifier of Insect Diversity (1745–1808)', in Robert Huxley (ed.), *The Great Naturalists* (London: Thames & Hudson, 2007), 182–85 and Mary Terrall, 'Following Insects Around: Tools and Techniques of Eighteenth-Century Natural History', *The British Journal for the History of Science,* 43 (2010), 573–88.

Research for this paper was kindly supported by a William Hunter Fellowship at the Hunterian, University of Glasgow and a DAAD/U4 Mobility grant at Uppsala Universitet in the summer of 2015. I would like to thank Andrew Wells and Sarah Cockram for invaluable comments on various drafts of this paper and the Enlightenment Reading Club, Göttingen for further help in improving the paper.

2   For eighteenth-century voyages of discovery, see John Gascoigne, *Encountering the Pacific in the Age of Enlightenment* (Cambridge: Cambridge University Press, 2014).

3   John Ellis to Carl Linnaeus, 19 Aug. 1768, The Linnaean correspondence, letter L4101 www.linnaeus.c18.net, accessed 12 July 2015. See Harold B. Carter, *Sir Joseph Banks, 1743–1820* (London: British Museum, 1988), 70–73 for further details on the instruments and containers aboard the HMS *Endeavour*.

4   Most notably Rob Iliffe, 'Science and Voyages of Discovery', in Roy Porter (ed.), *The Cambridge History of Science. The Eighteenth Century* (Cambridge: Cambridge University Press, 2003), 618–45 and David Philip Miller, 'Joseph Banks, Empire, and "Centers of Calculation" in Late Hanoverian London', in David Philip Miller and Peter Hanns Reill (eds), *Visions of Empire: Voyages, Botany and the Representations of Nature* (Cambridge: Cambridge University Press, 1996), 21–37. For eighteenth-century exploration voyages in general see Philip J. Stern, 'Exploration and Enlightenment', in Dane Kennedy (ed.), *Reinterpreting Exploration: The West in the World* (Oxford: Oxford University Press, 2014), 57–79.

5   Erica Fudge, 'What Was It Like to Be a Cow? History and Animal Studies', *Oxford Handbooks Online.* www.oxfordhandbooks.com/view/10.1093/oxfordhb/9780199927142.001.0001/oxfordhb-9780199927142-e-28, accessed 15 Aug. 2016.

6   Allan G. Gibbs, 'Water Balance in Desert Drosophila: Lessons from Non-Charismatic Microfauna', *Comparative Biochemistry and Physiology Part A,* 133 (2002), 781–89.

7   See Harry Liebersohn, 'A Half Century of Shifting Narrative Perspectives on Encounters', in Kennedy (ed.) *Reinterpreting Exploration*, 38–53 and Deepak Kumar, 'Botanical Explorations and the East India Company. Revisiting "Plant Colonialism"', in Vinita Damodaran, Anna Winterbottom and Alan Lester (eds) *The East India Company and the Natural World* (Basingstoke: Palgrave Macmillan 2015), 16–34.

8   See Paul Whitehead, 'Zoological Specimens from Captain Cook's Voyages', *Journal of the Society for the Bibliography of Natural History,* 5 (1969), 161–201 and Paul Whitehead, 'A Guide to the Dispersal of Zoological Material from Captain Cook's Voyages', *Pacific Studies,* 2 (1978), 52–93.

9   Søren L. Tuxen, 'Entomology Systematizes and Describes, 1700-1815', in Ray F. Smith, Thomas E. Mittler and Carol N. Smith (eds), *History of Entomology* (Palo Alto: Annual Reviews Inc., 1973), 95–118.

10 See e.g. Deirdre Coleman, 'Entertaining Entomology: Insects and Insect Performers in the Eighteenth Century', *Eighteenth-Century Life,* 30 (2006), 107–34; Elisabeth Wallmann, 'On Poets and Insects: Figures of the Human and Figures of the Insect in Pierre Perrin's *Divers Insectes* (1645)', *French History,* 28 (2014), 172–87; Pierre-Etienne Stockland, 'La Guerre Aux Insectes: Pest-Control and Agricultural Reform in the French Enlightenment', *Annals of Science,* 70 (2013), 435–60 and Mary Terrall, *Catching Nature in the Act: Réaumur and the Practice of Natural History in the Eighteenth Century* (Chicago: University of Chicago Press, 2014).

11 Michael S. Engel and Niels P. Kristensen, 'A History of Entomological Classification', *Annual Review of Entomology,* 58 (2013), 585–607 and Staffan Müller-Wille, 'Systems and How Linnaeus Looked at Them in Retrospect', *Annals of Science,* 70 (2013), 305–17.

12 Anne Mariss, *'A world of new things' Praktiken der Naturgeschichte bei Johann Reinhold Forster* (Frankfurt am Main: Campus, 2015), 181. See also David Elliston Allen, *The Naturalist in Britain. A Social History* (Princeton, NJ: Princeton University Press, 1976), 32–33 for earlier British manuals on collecting.

13 I am alluding to the term 'little tools of knowledge', with which Peter Becker and William Clark described written forms of knowledge making. I would however argue that object tools were as important in the process of knowledge production in natural history as were tables, maps and questionnaires. Cf. Peter Becker and William Clark, 'Introduction', in Becker and Clark (eds), *Little Tools of Knowledge. Historical Essays on Academic and Bureaucratic Practices* (Ann Arbor: University of Michigan Press, 2001), 1–34.

14 Johann Reinhold Forster, *A Catalogue of the Animals of North America* (London: B. White, 1771), 38.

15 See Georgina W. Brown, Starr Douglas and E. Geoffrey Hancock, 'The Use of Thorns and Spines as Pins in an Eighteenth Century Insect Collection', *The Linnean,* 27 (2011), 14–21 for further details on pinning techniques and the debate on using different materials in eighteenth century entomology.

16 Forster, *Catalogue*, 38. This method is apparently still in use in entomology to-day in the absence of a killing bottle. As insects have a ventral nerve cord this is instantaneous and arguably a quicker (and so less 'painful' or 'cruel') method of dispatch than in chemical fumes or by using the heat of the sun or fire. Personal communication, Geoff Hancock, Hunterian (Zoology) Museum, University of Glasgow, 10 Dec. 2015. I am immensely grateful to Geoff for sharing his knowledge of entomology and its history with me so generously.

17 Cf. however the clearly gendered attitudes towards insects in the early nineteenth century in Sam George, 'Animated Beings: Enlightenment Entomology for Girls', *Journal for Eighteenth-Century Studies,* 33 (2010), 487–505.

18 Forster, *Catalogue*, 38.

19 Sheila Wille, 'The Ichneumon Fly and the Equilibration of British Natural Economies in the Eighteenth Century', *British Journal for the History of Science,* 48 (2015), 639–60 at 648.

20 Johann Christian Fabricius, *Systema Entomologiae* (Flensburg and Leipzig, 1775) and Guillaume Antoine Olivier, *Entomologie, Ou Histoire Naturelle Des Insectes* (Paris, 1798). Cf. M. C. Day and Mike Fitton, 'Discovery in the Linnaean Collection of Type-Material of Insects Described by Johann Reinhold Forster, with Notes on the Hymenoptera', *Biological Journal of the Linnean Society,* 9 (1977), 31–43.

21 James Edward Smith, *The Natural History of the Rarer Lepidopterous Insects of Georgia* (London, 1797), p. i. Many thanks to Jeanne Robinson, Hunterian (Zoological) Museum, University of Glasgow for providing this reference and other valuable comments on a draft of this paper.

22 This contribution is repeated in a letter from 19 Jan. 1781, published in Neil Chambers (ed.), *Scientific Correspondence of Sir Joseph Banks, 1765–1820,* 6 vols (London: Pickering & Chatto, 2007), i. 260–62.

23 Pallas did indeed receive some plant seeds from Banks. See Folkwart Wendland, *Peter Simon Pallas (1741–1811): Materialien einer Biographie* (Berlin: Walter de Gruyter, 1992), 207.

24 See Simon Schaffer et al. (eds), *The Brokered World: Go-Betweens and Global Intelligence, 1770–1820* (Sagamore Beach: Science History Publications, 2009).

25 See Staffan Müller-Wille, 'Nature as a Marketplace. The Political Economy of Linnaean Botany', in Margaret Schabas and Neil De Marchi (eds), *Oeconomies in the Age of Newton* (Durham: Duke University Press, 2003), 154–72.

26 See Anke Te Heesen, 'From Natural Historical Investment to State Service: Collectors and Collections of the Berlin Society of Friends of Nature Research, c. 1800', *History of Science,* 42 (2004), 113–31.

27 Günter Bayerl and Jürgen Beckmann (eds), *Johann Beckmann (1739–1811). Beiträge zu Leben, Werk und Wirkung des Begründers der Allgemeinen Technologie* (Münster: Waxmann, 1999).

28 Johann Beckmann, 'Eine bequemere Einrichtung der Insektensammlungen', *Beschäftigungen der Berlinischen Gesellschaft Naturforschender Freunde, 2* (1776), 69–78. Beckmann uses the terms 'grausam' (cruel) and 'Widerwillen' (distaste).

29 Beckmann, 'bequemere Einrichtung', 77.

30 'Ich kann versichern, dass ich einige hundert Insekten, einige hundert Meilen weit, auf den unsanften Deutschen Postwagen, auf jene Weise verwahrt, unversehrt neben meinen Büchern und anderen Sachen transportiert habe'. Beckmann, *'bequemere Einrichtung'*, 78.

31 William Clark, Jan Golinski and Simon Schaffer, 'Introduction', in William Clark, Jan Golinski and Simon Schaffer (eds) *The Sciences in Enlightened Europe* (Chicago: The University of Chicago Press, 1999), 3–31 at 29.

32 David Philip Miller, 'Joseph Banks, Empire, and "Centers of Calculation" in Late Hanoverian London', in Miller and Reill (eds), *Visions of Empire*, 21–37.

33 See: Thomas Biskup, 'Transnational Careers in the Service of Empire: German Natural Historians in Eighteenth-Century London', in André Holenstein, Hubert Steinke and Martin Stuber (eds), *Scholars in Action: The Practice of Knowledge and the Figure of the Savant in the 18th Century* (Leiden: Brill, 2013), 45–69. Travel reports continued to be widely used however. See Bettina Dietz, 'Natural History as Compilation. Travel Accounts in the Epistemic Process of an Empirical Discipline', 703–19 in the same volume.

34 Johann Christian Fabricius, *Briefe aus London vermischten Inhalts* (Dessau und Leipzig, 1784), 1. In order to convince his superiors of the necessity to travel to Britain regularly, Fabricius also uses this argument in a letter to the German Chancellery in Copenhagen, from 16 Jan. 1777: 'England ist itzo das Land, welches die größten Entdeckungen in diesen Wissenschaften macht'. (England is now the country which makes the greatest discoveries in this science.) Landesarchiv Schleswig, Abt. 65.2, Nr. 561 II. See also Angus Armitage, 'A Naturalist's Vacation. The London Letters of J. C. Fabricius', *Annals of Science,* 14 (1958), 116–31.

35 Guillaume Antoine Olivier, *Entomologie, Ou Histoire Naturelle Des Insectes* (Paris, 1798), p. i: 'Qu'il me soit permis de témoigner ma reconnaissance aux Savans de Londres, qui ont bien voulu m'ouvrir leurs cabinets et leurs bibliothèques, aux Savans qui accueillent, avec bonté, les naturalistes étrangers qui se présentent chez eux, sans autre recommandation que leur zèle, sans autre titre que leur assiduité'. (Who will permit me to offer my gratitude to the learned people of London, who have kindly opened their cabinets and libraries to me, to those enlightened people who generously receive naturalists who present themselves with no other recommendation than their zeal and no other title than their assiduity.)

36 John Gascoigne, *Joseph Banks and the English Enlightenment. Useful Knowledge and Polite Culture* (Cambridge: Cambridge University Press, 2003), 70.

37 Letter of Joseph Banks to 'Count Lauragais', 6 Dec. 1772, in John Cawte Beaglehole (ed.), *The Endeavour Journal of Joseph Banks 1768–1771*, 2 vols (Sydney: Angus and Robertson Limited, 1962), ii. 328.

38 Indeed, when news arrived in continental Europe that a second circumnavigation was planned in 1771, Banks received a great number of letters asking for favours, specimens and employment. See John Cawte Beaglehole, *The Life of Captain James Cook* (Stanford: Stanford University Press, 1974), 291–92.

39 Letter of Johann Christian Fabricius to Joseph Banks, 29 Apr. 1776, in Chambers (ed.), *Scientific Correspondence*, i. 93–4 (no. 84).

40 Dru Drury, *Illustrations of Natural History* (London, 1770), p. vii.

41 Lisbet Koerner, *Linnaeus. Nature and Nation* (Cambridge: Harvard University Press, 2001).

42 James Boswell, *The Life of Samuel Johnson* (New York: Modern Library, 1952), 216, quoted by Roy Anthony Rauschenberg, 'Daniel Carl Solander: Naturalist on the Endeavour', *Transactions of the American Philosophical Society,* 58 (1968), 1–66.

43 Dru Drury, *Illustrations*, p. vi.

44 William Kirby and William Spence, *An Introduction to Entomology or Elements of the Natural History of the Insects* (London, 1818), p. v. Kirby and Spence also contend that the lack of textbooks in the vernacular hindered the development of British entomology. However, some of Kirby and Spence's predecessors in the eighteenth century had already published their accounts in English, like the multi-volume Edward Donovan, *The Natural History of British Insects* (London, 1792). Donovan also published on Australian, Chinese and Indian insects. Dru Drury had published his accounts bi-lingually in English and French. In his introduction Drury had already made the same observation as Kirby and Spence, namely that the study of nature, especially of insects, is frowned upon. Around 1800, it had become something of a topos for entomologists to lament the public contempt for their studies. Given the large number of actual publications and the private (and occasionally public) support for the studies of insects, this has to be taken with many grains of salt. Drury's engraver Moses Harris also published his work in English and French: Moses Harris, *The Aurelian or Natural History of English Insects* (London, 1778).

45 For the pre-Linnean history of entomology see Brian Ogilvie, 'The Pleasure of Describing: Art and Science in August Johann Rösel von Rosenhof's *Monthly Insect Entertainment*', in Liv Emma Thorsen, Karen A. Rader, and Adam Dodd (eds), *Animals on Display: The Creaturely in Museums, Zoos, and Natural History* (University Park: Penn State University Press, 2013), 77–100.

46 The following is based on E. Geoffrey Hancock, 'The Shaping Role of Johann Christian Fabricius: William Hunter's Insect Collection and Entomology in 18th-Century London', in E. Geoffrey Hancock, Mungo Campbell and Nick Pearce (eds), *William Hunter's World: The Art and Science of Eighteenth-Century Collecting* (Farnham: Ashgate, 2015), 151–63.

47 However, this new system was also inspired by Linnaeus, who hinted at this way of classification in 1740 in his second edition of his *Systema Naturae*, c.f. Søren L. Tuxen, 'Entomology Systematizes', 109.

48 Johann Christian Fabricius, 'Nova insectorum genera', *Skrivter af Naturhistorie-Selskabet,* 1 (1790), 213–28.

49 'Levemaade eller Oekonomie'. Fabricius, 'Nova insectorum genera', 214.

50 'og ofte i Sandhed vanskelige nok med Vished at kiende og at beskrive'. Fabricius, 'Nova insectorum genera', 214.

51 Adrian Hardy Haworth, 'Review of the Rise and Progress of the Science of Entomology in Great Britain; Chronologically Digested', *Transactions of the Entomological Society of London,* 1 (1812), 1–69 at 43.

52 'og disse Dele giemmer jeg bestandig til sammenligning'. Fabricius, 'Nova insectorum genera', 215.

53 Johann Karl Wilhelm Illiger, 'Ueber das Fabricische System und über die Bedürfnisse des jetzigen Zustandes der Insektenkunde', *Magazin für Insektenkunde,* 3 (1801), 261–84 at 269.

54 André Holenstein, Hubert Steinke and Martin Stuber, 'Introduction', in André Holenstein, Hubert Steinke and Martin Stuber (eds), *Scholars in Action: The Practice of Knowledge and the Figure of the Savant in the 18th Century* (Leiden: Brill, 2013), 1–41.

55 'Selbst manche Frauenzimmer haben es in diesen Untersuchungen und Zubereitungen sehr weit gebracht'. Johann Christian Fabricius, 'Vertheidigung des Fabricischen Systems', *Magazin für Insektenkunde,* 2 (1803), 1–13 at 6.
56 Frank N. Egerton, *Roots of Ecology: Antiquity to Haeckel* (Berkeley: University of California Press, 2012).
57 'welche die Insektenlehre nicht als Wissenschaft, sondern als Liebhaberei betreiben'. Fabricius, 'Vertheidigung', 11.

Bibliography

Allen, David Elliston, *The Naturalist in Britain. A Social History* (Princeton: Princeton University Press, 1976).

Armitage, Angus, 'A Naturalist's Vacation. The London Letters of J. C. Fabricius', *Annals of Science,* 14 (1958), 116–31.

Bayerl, Günter and Jürgen Beckmann (eds), *Johann Beckmann (1739–1811). Beiträge zu Leben, Werk und Wirkung des Begründers der Allgemeinen Technologie* (Münster: Waxmann, 1999).

Beaglehole, John Cawte (ed.), *The Endeavour Journal of Joseph Banks 1768–1771,* 2 vols (Sydney: Angus and Robertson Limited, 1962).

———, *The Life of Captain James Cook* (Stanford: Stanford University Press, 1974).

Becker, Peter and William Clark, 'Introduction', in Peter Becker and William Clark (eds), *Little Tools of Knowledge. Historical Essays on Academic and Bureaucratic Practices* (Ann Arbor: University of Michigan Press, 2001), 1–34.

——— (eds), *Little Tools of Knowledge. Historical Essays on Academic and Bureaucratic Practices* (Ann Arbor: University of Michigan Press, 2001).

Beckmann, Johann, 'Eine bequemere Einrichtung der Insektensammlungen', *Beschäftigungen der Berlinischen Gesellschaft Naturforschender Freunde,* 2 (1776), 69–78.

Biskup, Thomas, 'Transnational Careers in the Service of Empire: German Natural Historians in Eighteenth-Century London', in André Holenstein, Hubert Steinke and Martin Stuber (eds), *Scholars in Action: The Practice of Knowledge and the Figure of the Savant in the 18th Century* (Leiden: Brill, 2013), 45–69.

Brown, Georgina W., Starr Douglas and E. Geoffrey Hancock, 'The Use of Thorns and Spines as Pins in an Eighteenth Century Insect Collection', *The Linnean,* 27 (2011), 14–21.

Carter, Harold B., *Sir Joseph Banks, 1743–1820* (London: British Museum [Natural History], 1988).

Chambers, Neil (ed.), *Scientific Correspondence of Sir Joseph Banks, 1765–1820,* 6 vols (London: Pickering & Chatto, 2007).

Clark, William, Jan Golinski and Simon Schaffer, 'Introduction', in William Clark, Jan Golinski and Simon Schaffer (eds), *Sciences in Enlightened Europe* (Chicago: The University of Chicago Press, 1999), 3–31.

——— (eds), *The Sciences in Enlightened Europe* (Chicago: University of Chicago Press, 1999).

Coleman, Deirdre, 'Entertaining Entomology: Insects and Insect Performers in the Eighteenth Century', *Eighteenth-Century Life,* 30 (2006), 107–34.

Damodaran, Vinita, Anna Winterbottom and Alan Lester (eds), *The East India Company and the Natural World* (Basingstoke: Palgrave Macmillan 2015).

Day, M. C. and Mike Fitton, 'Discovery in the Linnaean Collection of Type-Material of Insects Described by Johann Reinhold Forster, with Notes on the Hymenoptera', *Biological Journal of the Linnean Society,* 9 (1977), 31–43.

Dietz, Bettina, 'Natural History as Compilation. Travel Accounts in the Epistemic Process of an Empirical Discipline', in Holenstein, Steinke and Stuber (eds), *Scholars in Action*, 703–19.

Donovan, Edward, *The Natural History of British Insects* (London, 1792).

Drury, Dru, *Illustrations of Natural History* (London, 1770).

Egerton, Frank N., *Roots of Ecology: Antiquity to Haeckel* (Berkeley: University of California Press, 2012).

Engel, Michael S. and Niels P. Kristensen, 'A History of Entomological Classification', *Annual Review of Entomology,* 58 (2013), 585–607

Fabricius, Johann Christian, *Systema Entomologiae* (Flensburg and Leipzig, 1775).

———, *Briefe aus London vermischten Inhalts* (Dessau und Leipzig, 1784).

———, 'Nova insectorum genera', *Skrivter af Naturhistorie-Selskabet,* 1 (1790), 213–28.

———, 'Vertheidigung des Fabricischen Systems', *Magazin für Insektenkunde,* 2 (1803), 1–13.

Forster, Johann Reinhold, *A Catalogue of the Animals of North America* (London, 1771).

Fudge, Erica, 'What Was It like to Be a Cow? History and Animal Studies', *Oxford Handbooks Online,* www.oxfordhandbooks.com/view/10.1093/oxfordhb/9780199927142.001.0001/oxfordhb-9780199927142-e-28, accessed 15 Aug. 2016.

Gascoigne, John, *Joseph Banks and the English Enlightenment. Useful Knowledge and Polite Culture* (Cambridge: Cambridge University Press, 2003).

———, *Encountering the Pacific in the Age of Enlightenment* (Cambridge: Cambridge University Press, 2014).

George, Sam, 'Animated Beings: Enlightenment Entomology for Girls', *Journal for Eighteenth-Century Studies,* 33 (2010), 487–505.

Gibbs, Allan G., 'Water Balance in Desert Drosophila: Lessons from Non-Charismatic Microfauna', *Comparative Biochemistry and Physiology Part A,* 133 (2002), 781–89.

Hancock, E. Geoffrey, Mungo Campbell and Nick Pearce, 'The Shaping Role of Johann Christian Fabricius: William Hunter's Insect Collection and Entomology in 18th-Century London', in E. Geoffrey Hancock, Mungo Campbell and Nick Pearce (eds), *William Hunter's World: The Art and Science of Eighteenth-Century Collecting* (Farnham: Ashgate, 2015), 151–63.

——— (eds), *William Hunter's World: The Art and Science of Eighteenth-Century Collecting* (Farnham: Ashgate, 2015).

Harris, Moses, *The Aurelian or Natural History of English Insects* (London, 1778).

Haworth, Adrian Hardy, 'Review of the Rise and Progress of the Science of Entomology in Great Britain; Chronologically Digested', *Transactions of the Entomological Society of London,* 1 (1812), 1–69.

Holenstein, André, Hubert Steinke and Martin Stuber, 'Introduction', in André Holenstein, Hubert Steinke and Martin Stuber (eds), *Scholars in Action: The Practice of Knowledge and the Figure of the Savant in the 18th Century* (Leiden: Brill, 2013), 1–41.

——— (eds), *Scholars in Action: The Practice of Knowledge and the Figure of the Savant in the 18th Century* (Leiden: Brill, 2013).

Huxley, Robert (ed.), *The Great Naturalists* (London: Thames & Hudson, 2007).

Iliffe, Rob, 'Science and Voyages of Discovery', in Roy Porter (ed.), *Cambridge History of Science. The Eighteenth Century* (Cambridge: Cambridge University Press, 2003), 618–45.

Illiger, Johann Karl Wilhelm, 'Ueber das Fabricische System und über die Bedürfnisse des jetzigen Zustandes der Insektenkunde', *Magazin für Insektenkunde,* 3 (1801), 261–84.

Kennedy, Dane (ed.) *Reinterpreting Exploration: The West in the World* (Oxford: Oxford University Press, 2014).

Kirby, William and William Spence, *An Introduction to Entomology or Elements of the Natural History of the Insects* (London, 1818).

Koerner, Lisbet, *Linnaeus. Nature and Nation* (Cambridge: Harvard University Press, 2001).

Kumar, Deepak, 'Botanical Explorations and the East India Company. Revisiting 'Plant Colonialism', in Vinita Damodaran, Anna Winterbottom and Alan Lester (eds), *The East India Company and the Natural World* (Basingstoke: Palgrave Macmillan 2015), 16–34.

Liebersohn, Harry, 'A Half Century of Shifting Narrative Perspectives on Encounters', in Kennedy (ed.) *Reinterpreting Exploration*, 38–53.

Mariss, Anne, *'A World of New Things': Praktiken der Naturgeschichte bei Johann Reinhold Forster* (Frankfurt am Main: Campus, 2015).

Miller, David Philip and Peter Hanns Reill, 'Joseph Banks, Empire, and "Centers of Calculation" in Late Hanoverian London', in David Philip Miller and Peter Hanns Reill (eds), *Visions of Empire: Voyages, Botany and the Representations of Nature* (Cambridge: Cambridge University Press, 1996), 21–37.

——— (eds), *Visions of Empire: Voyages, Botany and the Representations of Nature* (Cambridge: Cambridge University Press, 1996).

Müller-Wille, Staffan, 'Nature as a Marketplace. The Political Economy of Linnaean Botany', in Margaret Schabas and Neil De Marchi (eds), *Oeconomies in the Age of Newton* (Durham: Duke University Press, 2003), 154–72.

———, 'Systems and How Linnaeus Looked at Them in Retrospect', *Annals of Science,* 70 (2013), 305–17.

Ogilvie, Brian, 'The Pleasure of Describing: Art and Science in August Johann Rösel von Rosenhof's *Monthly Insect Entertainment*', in Liv Emma Thorsen, Karen A. Rader and Adam Dodd (eds), *Animals on Display: The Creaturely in Museums, Zoos, and Natural History* (University Park: Penn State University Press, 2013), 77–100.

Olivier, Guillaume Antoine, *Entomologie, Ou Histoire Naturelle Des Insectes* (Paris, 1798).

Porter, Roy (ed.), *The Cambridge History of Science. The Eighteenth Century* (Cambridge: Cambridge University Press, 2003).

Rauschenberg, Roy Anthony, 'Daniel Carl Solander: Naturalist on the Endeavour', *Transactions of the American Philosophical Society,* 58 (1968), 1–66.

Schabas, Margaret and Neil De Marchi (eds), *Oeconomies in the Age of Newton* (Durham: Duke University Press, 2003).

Schaffer, Simon et al. (eds), *The Brokered World: Go-Betweens and Global Intelligence, 1770–1820* (Sagamore Beach: Science History Publications, 2009).

Smith, James Edward, *The Natural History of the Rarer Lepidopterous Insects of Georgia* (London, 1797).

Smith, Ray F., Thomas E. Mittler and Carol N. Smith (eds), *History of Entomology* (Palo Alto: Annual Reviews, 1973).

Stern, Philip J., 'Exploration and Enlightenment', in Dane Kennedy (ed.), *Reinterpreting Exploration: The West in the World* (Oxford: Oxford University Press, 2014), 57–79.

Stockland, Pierre-Etienne, 'La Guerre Aux Insectes: Pest-Control and Agricultural Reform in the French Enlightenment', *Annals of Science,* 70 (2013), 435–60.

Te Heesen, Anke, 'From Natural Historical Investment to State Service: Collectors and Collections of the Berlin Society of Friends of Nature Research, c. 1800', *History of Science,* 42 (2004), 113–31.

Terrall, Mary, *Catching Nature in the Act: Réaumur and the Practice of Natural History in the Eighteenth Century* (Chicago: University of Chicago Press, 2014).

———, 'Following Insects Around. Tools and Techniques of Eighteenth-Century Natural History', *The British Journal for the History of Science,* 43 (2010), 573–588.

Thorsen, Liv Emma, Karen A. Rader and Adam Dodd (eds), *Animals on Display. The Creaturely in Museums, Zoos, and Natural History* (University Park: Penn State University Press, 2013).

Tuxen, Søren L., 'Entomology Systematizes and Describes, 1700–1815', in Ray F. Smith, Thomas E. Mittler and Carol N. Smith (eds), *History of Entomology* (Palo Alto: Annual Reviews Inc., 1973), 95–118.

Vane-Wright, Richard I., 'Johann Christian Fabricius. Classifier of Insect Diversity (1745–1808)', in Robert Huxley (ed.), *Great Naturalists* (London: Thames & Hudson, 2007), 182–5.

Wallmann, Elisabeth, 'On Poets and Insects: Figures of the Human and Figures of the Insect in Pierre Perrin's *Divers Insectes* (1645)', *French History,* 28 (2014), 172–87.

Wendland, Folkwart, *Peter Simon Pallas (1741–1811): Materialien einer Biographie* (Berlin: Walter de Gruyter 1992).

Whitehead, Paul, 'A Guide to the Dispersal of Zoological Material from Captain Cook's Voyages', *Pacific Studies,* 2 (1978), 52–93.

———, 'Zoological Specimens from Captain Cook's Voyages', *Journal of the Society for the Bibliography of Natural History,* 5 (1969), 161–201.

Wille, Sheila, 'The Ichneumon Fly and the Equilibration of British Natural Economies in the Eighteenth Century', *British Journal for the History of Science,* 48 (2015), 639–60.

9 Hungarian grey cattle

Parallels in constituting animal and human identities

László Bartosiewicz

In other parts of Europe there are several distinct races, such as the pale-coloured Hungarian cattle, with their light and free step, and their enormous horns sometimes measuring above five feet from tip to tip.

—Charles Darwin (1868)[1]

Introduction

Ideas of nation and nationhood in the modern world are the product of a number of forces, currents, and trends, whether political, economic, social, intellectual, cultural, or natural. Often overlooked as an influence shaping national, and more broadly human, identities are animals. One need look no further than national fauna – the British lion, French cock, or Russian bear, for instance – to recognize one aspect of how animals help to shape human identities. Nor is this process entirely one-way, for while modern concepts of nations were being produced during the Age of Enlightenment, so too were tightly defined notions of animal breeds.

Nations and breeds had more in common than mere simultaneity. In a modern sense, 'breed' denotes a relatively homogenous group of animals, developed and maintained by humans, frequently on a territorial basis. A homogeneous group occupying a common territory has been one influential way of defining the nation, and although it is perhaps more common in the breach than the observance, this definition has nevertheless often been at the heart of national self-perception. Today the terms 'breed' and 'nation' are frequently applied anachronistically to ancient periods before they had crystallized into their present-day forms. Animals of a common breed would presumably not be conscious of their unity, but through their human-driven development they became more than passive objects of human categorization impacting on culture.

In this chapter, I argue that the formation of national identities and modern breeds were intertwined and mutually dependent processes and that sharing territory was fundamental for both. The strength of this connection should not be overstated, but there nonetheless exists a remarkable homology between the way breeds and nations have been seen in retrospect.

Live domestic animals are bona fide artefacts of human culture as they have been, both consciously and unconsciously, shaped by people not simply to meet material needs but also to express tastes, aspirations, and various forms of social identity. Importantly, however, this relationship is not unidirectional. Domesticates are not merely static reflections of culture: in ever-changing natural and social environments, they influence and mediate ideas as well as evoke an array of emotions. As a result, animals have the potential to modify human social behaviour through a range of often-subtle feedback effects.

Animals even have the potential to influence human culture from beyond the grave. The breed that forms the case study explored in this essay, Hungarian Grey cattle, was used as a potent symbol and key plank of narratives of 'national' origins stretching over a millennium. But such narratives lack congruent osteological evidence. Notwithstanding romantic mythology, the earliest relevant archaeological finds are contemporaneous with the early modern imagery of these animals. The earliest specimens date from the tail end of the post-medieval era in which the long-distance livestock trade flourished. The human context was also variable, as both sourcing areas in the periphery of Europe and destination markets in its urbanized core were constantly shifting as the continent underwent radical geopolitical changes that over the long run resulted in the emergence of nation states.

The discussion will address several key themes to explore the relationship of Hungarian Grey cattle and Hungarian nationhood. The following section explores the importance of appearance and the salient characteristics of the breed. By the nineteenth century breeds were established (and have since largely been cultivated) with a strong emphasis on external features. Trademark appearances have served to guarantee the presence of production traits associated with them, but they also tend to have an emotional impact on the onlooker. Aside from the direct visual effect, narratives concerning the origins of breeds are part of a symbolic identity that may thus be manipulated. The chapter then proceeds to examine three different theories that have been offered to account for the origins of the Hungarian Grey. Each of these has been used to serve particular ideological and mythological agendas but, as the discussion will move to point out, zooarchaeological support has been fundamentally lacking. Two further sections will explore the parallel histories of Hungarian nationhood and the Hungarian Grey from the seventeenth to twentieth centuries, to demonstrate the complexity of the relationship of nation and breed.

Keeping up appearances

For millennia, cattle have been important sources of beef and numerous renewable products such as milk, traction, and manure. Much of traditional animal breeding revolved around appearances: the exterior of livestock was evident and easily judged by the expert eye. Why is their mere appearance so important? Beyond the symbolism and aesthetics that are at the heart of the 'artefactual' aspect of livestock, appearances are known to be a driving force behind the conscious breeding of animals. References to the beauty of domestic animals in surviving documents point not merely to the nature of the animals but to the breeding skills, trade connections, or at least the wealth and status of their owners. Beyond this tendency to objectify animals as products, the appearances of animals are also inseparable from stereotypes of behaviour sought by the breeders and owners. Alongside sheer strength, docility, and predictability are practical virtues in working animals (traditionally improved by early castration).[2]

Symbolic imagery, however, may override down-to-earth technical advantages. In the case of bulls, such symbolism has often related to strength and male virility. Bull fighting and wrangling have been traditional tests of courage in numerous cultures. Middle Bronze Age (*c.* fourteenth century BCE) bull-leaping (*taurokathapsia*) is best known from Minoan Crete, but was also found throughout Southwest Asia as well as ancient Egypt.[3] Contemporary bull-leaping (*course landaise*) contests are still organized in Gascony in south-west France. Cows are increasingly used as substitutes for bulls in modern games of this sort because they are less dangerous (Figure 9.1), and in many societies they have traditionally embodied less combative and more feminine characteristics, such as docility and nourishment. The Egyptian goddess Hathor frequently appears in the guise of a cow suckling the pharaoh, for example in the eighteenth-dynasty (mid-fifteenth century BCE) temple of Djeser Akhet.[4] In the *Iliad*, Hera is described by Homer as 'cow-eyed'.[5]

Figure 9.1 Cowherds' competition in Hungary, 2012 (photo: Máté Babay).

In addition to this gendered dichotomy, relations between animals and humans are often defined according to other dualistic perceptions of 'good' and 'bad', 'useful' and 'dangerous' and so on for most species.

The breed that Darwin refers to as 'pale-coloured' is known today as the Hungarian Grey cattle,[6] a distinctive strain whose additional colour varieties in the vernacular nomenclature include ashy, blue-roan, crane-like, crane-coloured, cream-coloured, dark crane, dusty, frosty, jackdaw, silvery white, silvery grey, slate-grey, smoky, sooty, and thrush-coloured.[7] Reddish shades may occur in the winter coat, and calves are invariably born red, only turning grey by the age of six months.[8]

In addition to colour, particular attention has been devoted to horn conformation in these animals. Ottó Herman, the early twentieth-century polymath, collected no fewer than 172 terms referring to horn shape and direction in cattle.[9] Long, symmetric, and widely set horns (i.e. separated by a broad intercornual ridge) have traditionally been associated with a desirable constitution.[10] Horn length must exceed the length of the head by at least 50 percent in cows, while in bulls horns should be at least as long as the head according to the breed's standard.[11]

This standard was compiled by the late twentieth century, based on detailed studies of breeding documentation reaching back to 1937.[12] Measuring the importance of various traits by the number of words dedicated to major external features, this official definition of Hungarian Grey cattle may be compared in interesting ways to the 1885 standard of Scottish Highland cattle, another iconic traditional breed.[13] In both cases, external features, especially hair colour and horn conformation, are described in far more detail than the animals' body shape. While in terms of absolute word count the head is discussed to a comparable extent in the more laconically written Highland cattle standard, two thirds of the description are dedicated to colour and horn shape in the case of Hungarian Grey (Figure 9.2).

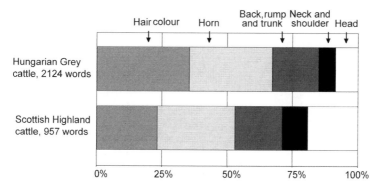

Figure 9.2 The number of words devoted to the description of external appearance in the standards of Hungarian grey and Scottish highland cattle (author's drawing).

The call of the wild: aurochs origins?

Establishing the appearance and characteristics of the Hungarian Grey is of paramount importance in the effort to account for its origins. Three key theories have attempted to provide such an explanation and have focused on the possibility of descent from the ancient aurochs, from cattle brought by eastern migrants to medieval Hungary, and from livestock introduced during the early modern heyday of the long-distance cattle trade. Each of these theories served specific ideological and mythological goals, as indeed attitudes towards animals in general have frequently been manipulated to convey symbolic messages.

A pertinent example of animal breeding serves to illustrate broadly how the perception of a newly 'created' domestic animal can evolve within a radicalizing ideological context. In 1920, the Heck brothers, Lutz (1892–1983, director of the Zoologischer Garten in Berlin) and Heinz (1894–1982, director of the Tierpark Hellabrunn in Munich) began attempting to reconstruct the extinct aurochs by crossing traditional breeds with aurochs-like phenotypic traits. Heinz included the Hungarian Grey in his work in Munich, while Lutz used another stock of unimproved cattle breeds in Berlin.[14] The aurochs has for centuries been conflated with European bison.[15] The Heck brothers originally intended to exhibit both species side by side for educational purposes. Although this long-term project was initiated in the Weimar Republic, Lutz Heck published its first results the year after Hitler was sworn in as chancellor.[16] Subsequent sponsorship of continued research by *Reichsjägermeister* (Hunt Master of the Empire) Hermann Göring tarnished the reputation of these animals in the popular media as 'specifically bred by Nazi scientists'.[17] This characterisation of their work is perhaps unfair, but the Heck brothers undeniably swam with the prevailing tide. The resulting animals were not, however, aurochs but a new breed of domestic cattle carefully designed to *look* similar to the wild ancestor, hence their official name: Heck cattle (*Heckrind*). The danger posed to people by these beasts may be attributed as much to the way in which they have been habitually kept as free-ranging animals as to their genetic makeup. Their ferocity, however, has recently evoked memories of sinister Nazi patronage during the second half of the project.[18]

Anachronism is another prominent feature of the Heck experiment. Evolution is permanent and irreversible: the ancestors or even past breeds of domesticates cannot be genetically reconstructed; even maintaining an existing breed is a dynamic process requiring active intervention. Attempts to breed humans back to 'primeval Aryan stock' in the Third Reich shows how rigid concepts of eugenics were connected to the ideals of animal breeding programmes.[19] Beyond the guiding Nazi principle of racial purity, the back-crossed 'aurochs', concocted from a host of domestic breeds, ended up contributing to an ill-fated propaganda environment that enhanced its own symbolic content. The paradigmatic story of Heck cattle shows not only rapidly shifting criteria in the perception of, and attitudes toward, animals under changing historical circumstances. It also demonstrates the inherent mutualism of human-animal interactions.

The Hecks, moreover, were not working in a vacuum. The idea that the Hungarian Grey descended directly from the aurochs (*Bos primigenius*, Bojanus 1827, the now extinct ancestor of domestic cattle) was considered from the mid-nineteenth century by the prominent Swiss archaeozoologist, Ludwig Rütimeyer, among other scholars.[20] This notion was also entertained by Charles Darwin who greatly respected Rütimeyer's pioneering osteological work on the history of domestic animals.[21] Perhaps implausibly, other scholars, while arguing that the Hungarian Grey descended from the ancient aurochs, also claimed that it was an autochthonous breed, on the basis that local domestication is the only legitimate foundation for calling a breed 'autochthonous'. In the mid-twentieth century, for example, Miklós Jankovich postulated that the Hungarian Grey was domesticated in medieval Hungary by the '*venatores bubalinorum*' ('bubalus-calf hunters') mentioned in a thirteenth-century document.[22]

The significance of aurochs ancestry is related to the image of Hungarian Grey cattle as hardy, often fierce beasts, an image that served specific ideological goals. A romantic fascination with the 'wild' and with taming savage beasts as a token of male prowess is a well-known cultural phenomenon and has proven to be a persistent feature of the traditional image of the Hungarian Grey cattle irrespective of political context. During the years of Stalinist rule in the early 1950s, a new image of the 'people' (meaning working classes rather than *Volk*) was consolidated in Hungary.[23] A romanticized past is beautifully reflected on a 1955 porcelain statuette that shows the medieval folk hero, Miklós Toldi, overcoming a Hungarian Grey bull (Figure 9.3).

Figure 9.3 Herend porcelain statuette designed by János Tóth, 1955, Galéria Savaria online piactér.

Zooarchaeology has not smiled on the romantic idea of direct descent from the aurochs. It has been long noted that by the thirteenth century aurochs bones become extremely rare in archaeological assemblages from Hungary; extinction proceeded largely along a south–north axis in Europe.[24] As once recognized by the Heck brothers, early modern references to wild cattle are more likely to have concerned European bison that survived in hilly areas of Hungary until the early nineteenth century.[25] Meanwhile, osteometric data show that most early medieval cattle in the Carpathian Basin were unusually small (withers height *c.*110 cm), short-horned (*brachyceros* type) animals with no phenotypes of transitional sizes that would be potentially indicative of on-going domestication.[26] During the critical medieval period, neither the wild ancestor nor the resulting impressive breed can be sufficiently documented. The likelihood of local domestication was further refuted by mitochondrial DNA evidence indicative of the Near Eastern origins of domestic cattle with no convincing genetic links to the local European aurochs.[27]

Ex oriente lux: introduction by migrating peoples?

Another (and probably the most) popular myth concerning the origins of the Hungarian Grey is that it was brought along by conquering Hungarians from the Eurasian steppe belt. This version of *ex oriente lux* – the theory that innovation in Europe was generated by impulses from the East – is at least as old as the idea of local domestication. In a significant number of nineteenth-century romantic representations, the breed is portrayed as a great contribution to European culture by the conquering Hungarians who occupied the Carpathian Basin in the ninth century and raided neighbouring territories for at least another hundred years.[28] The misconception that 'autochthonous' Hungarian Grey cattle were introduced by 'nomadic' ancient Hungarians gained a special impetus in the newly established Austro-Hungarian monarchy when Hungarian national identity was asserted at all levels.[29] Celebration of the 1896 millennium of the Hungarian Conquest marked a political climax after the 1867 *Ausgleich* (compromise) with Austria, mustering the achievements of 1000 years by the Hungarian 'nation'. One piece of art to commemorate this event is the panoramic painting by Árpád Feszty, a perfect example of political propaganda wrapped in the packaging of romantic historicism.[30] Feszty's magnificent imagery has been viewed for well over a century as typical of nineteenth-century historical stereotypes. Among others, he depicted the most beautiful Hungarian Grey oxen of his time (Figure 9.4), illustrating the state of historical and archaeological knowledge as a curious monument to the mentality of an era.

Variants of this myth also exist. The belief in Hungarian origin has ironically served other accounts, in part because of the perpetual confusion between fifth-century Huns and ninth-century Hungarians in both historical sources and popular literature. Consequently, this myth also contributed to the naïve hypothesis that long-horned, grey Maremman cattle were

Figure 9.4 The romantic depiction of 'Conquest Period' Hungarian grey oxen by Árpád Feszty. (Detail from engraving by Gusztáv Morelli, 1895).

introduced to Italy by Attila's advancing army, an idea reinforced by his amply documented encounter with Pope Leo I in 452.[31] Another possibility was suggested in the 1970s by János Matolcsi, who argued that Hungarian Grey cattle may have been introduced during the last waves of eastern migrations represented by Cumanians in the thirteenth to fifteenth centuries.[32]

Archaeological evidence, however, no more supports this cluster of myths than it does those linking the Hungarian Grey to the aurochs. Excavations have not yielded evidence of long-horned cattle from the most thoroughly investigated sixteenth-century Cumanian settlement of Szentkirály or any other contemporaneous site.[33] In addition, coeval written references to such animals are missing. There is, furthermore, a serious gap in the archaeozoological record directly relating to the breed, which achieved decisive economic importance between the eighteenth and early twentieth centuries. The period after 1711, when the Szatmár Peace Accord was concluded between the Habsburgs and Hungary, is not officially considered to have archaeological significance in Hungary.[34] Hungarian Greys are not studied by conducting targeted excavations, and accidental finds are hardly ever collected during archaeological fieldwork.

Aside from the much-discussed arrival of Hungarians with their livestock of presumably steppe origins, subsequent external influences on these animals have rarely been discussed. The relationship between Hungarian Grey and similar breeds in Italy is still poorly understood.[35] Commercial imports of this type of cattle from the Neapolitan court during fourteenth-century Anjou rule in Hungary have also been hypothesized.[36] Indeed, during the rule of the Habsburg Grand Dukes of Tuscany (1737–1859), Maremman sires were reputedly imported to Hungary to improve local stocks. Knowledge of this led in the early 1970s to the introduction of three Maremman bulls to ease the genetic bottleneck effect in the dramatically reduced population.[37]

Cattle trade

The final theory concerning the origin of the Hungarian Grey to be considered here is related to the long-distance livestock trade. This was of critical importance in the Hungarian economy between the sixteenth and eighteenth centuries. As in other areas peripheral to the increasingly urbanized core of Central and Western Europe, extensive grazing in the hinterlands made transcontinental meat provisioning for urban markets a lucrative enterprise. Livestock needs room to rest, food, and water. The logistics of this enterprise had to be professionally organized and bankrolled. An amply documented large-scale cattle trade in the broader region of medieval Hungary meant that every year hundreds of thousands of animals were herded to feed growing cities in northern Italy and southern Germany.[38] In turn, Hungarians imported industrial products: craft and trade goods such as fine textiles, high quality metal tools and weapons.

According to Sándor Takáts, a 1526 document mentions 'Hungarian cattle' having been introduced to the Augsburg market, although written evidence for westward exports is known from as early as the mid-fourteenth century.[39] A chance explicit reference to 'long-horned Hungarian cattle' (*magnus cornuotes boves Hungaricos*), however, first appears only in a late sixteenth-century document when exports from Hungary indeed peaked.[40] According to the Ottoman Turkish tax rolls from Vác, a market town on the Danube in Hungary, herd sizes varied greatly although on average at least a dozen groups of 100–120 cattle grazed their way toward the west.[41] The journey took three to four months covering up to a thousand kilometres to the markets. This relatively slow pace (8–12 km per day) would have been necessary to prevent the animals from becoming too lean during the lengthy trek, in order to secure the best price at market. The animals were accompanied by half a dozen 'ox captains' per hundred beasts. As has been widely illustrated in art depicting the American West, this was a strongly masculine and often violent world. In times of warfare, herders are known to have formed paramilitary units, and they provided the core of *heyducks*, sixteenth-century mercenaries engaged in irregular warfare against both the Turks and Habsburgs, as well as opportunistic brigandage.[42]

Although it would be erroneous to speak of a breed at this early date, the nineteenth-century romantic mythology of hardy, sturdy Hungarian Grey cattle that could be tamed only by men equally robust is probably rooted in this period. These animals exerted a feedback effect on sixteenth-century society through large-scale trade, stimulating the emergence of entrepreneurs and their range of employees, some of whom served as irregulars in conflict situations. Sixteenth-century westward drives originated from redistribution

Figure 9.5 Stone ox decorating the Fleischbrücke in Nürnberg (author's photo).

centres near the Danube Bend, trekked past Bratislava, Vienna, and St. Pölten, crossed the Inn at Schärding (by swimming), moved on to Regensburg, and continued toward Nuremberg or Augsburg.[43]

This myth also left architectural traces. Upon the completion of the famous late Renaissance *Fleischbrücke* in Nuremberg in 1598, a side gate was added leading to the city's meat hall, decorated with a long-horned stone ox (Figure 9.5). Due to its striking similarity, many commentators have been practically obsessed with the idea that this statue was modelled after a Hungarian Grey, which supposedly made up the majority of cattle exported during the period of Hungarian hegemony over livestock markets south of the Main River (1470–1574).[44]

Osteological evidence and imagery

The emergence of the Hungarian Grey from the cattle trade seems the most plausible account of its origins, but all theories have to face the challenge posed by archaeological evidence. Although it is extremely tempting to equate medieval domesticates with modern breeds, bone remains suggest that the familiar appearance of our domestic animals became consolidated only during the last two to three hundred years.[45] Any static view of breeds disregards two indisputable facts: breeds continue to develop in a biological sense under human influence, and the concept of breed itself has evolved culturally as generations of animals contributed to it.

Reliably dated horn cores of unusually large size would provide the only osteological proof for early Hungarian Grey cattle. A fatal flaw of theories emphasising the medieval origins of the breed is that none of them is supported by such tangible material evidence, despite the tens of thousands of medieval cattle bones that have been regularly identified for over six decades by dozens of experts in Hungary. The horn sheath itself, covering the bony horn core, was a valuable raw material throughout history. Extremely large horns could also have been set aside as decoration or 'trophies'. Such finds, however, begin to surface only from the Early Modern Age onward in Hungary. To date, the longest intact horn core finds have been the three specimens recovered from a fourteenth- to seventeenth-century deposit in the city of Kecskemét at the site of Bocskai *utca* (street). The largest of these was 427 mm long (length of the outer curvature), and the circumference of its base was 266 mm.[46] The lengths of six measurable horn cores from an eighteenth-century industrial pit in Budapest–Szalag *utca* fell between 300 and 400 mm.[47] Although some of the horn sheaths on these cores could have been near 50 cm long, they were significantly shorter than those of present-day Hungarian Grey, and their shapes also appear different.[48] On the other hand, there is a remarkable similarity between the dimensions of the eighteenth-century horn cores and those of modern Hungarian Grey cattle (Figure 9.6). But it must be noted that not even long horn cores from archaeological sites can be directly equated with the breeds known today.

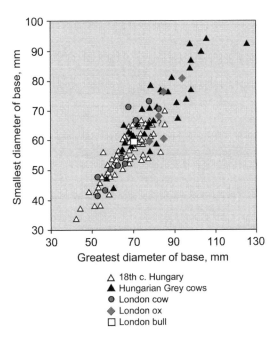

△ 18th c. Hungary
▲ Hungarian Grey cows
● London cow
◆ London ox
□ London bull

Figure 9.6 Comparison of horn core base diameters between eighteenth-century finds from Hungary, present-day Hungarian grey cows, and horn cores of adult seventeenth- and eighteenth-century cattle from London (author's drawing, based on Csippán 2009 and Armitage 1982).

Most cattle bones from excavations at key settlements in Hungary involved in livestock provisioning to markets in Central Europe originate from a relatively non-characteristic admixture of forms. Naturally, not all animals slaughtered at these sites originated from cattle drives, but animals taken to markets were slaughtered along the way. Trading posts also benefited at least seasonally from a better supply of beef than ordinary settlements. With the passage of time, demand increased for large and leggy animals, to both provide quantities of meat and be fit enough to cover long distances. First size variability, then an increase of stature can be detected in archaeological bone measurements.[49] The complete lack of medieval archaeozoological evidence for long horns, however, suggests that the emergence and gradual homogenization of external features in the Hungarian Grey cattle were neither as ancient nor as sudden as nineteenth- and twentieth-century mythologies would suggest. Moreover, mid-seventeenth-century records mention a whole range of cattle colours, including blonde, piebald, yellow, black, blue, and shades of brown.[50] It is easy to imagine that given the profitability of cattle drives, animals of all sorts were mobilized.

It was during these years that large, long-horned cattle became fashionable across Europe. In Hungary, sporadic fragments of large eighteenth-century horn cores were also reported from the city of Vác, where a contemporaneous Baroque column head displays long-horned cattle.[51] Such cattle were already present in early modern Britain. The English Longhorn became the first breed to be systematically improved for beef production by Robert Bakewell (1725–1795), when demand for quantities of meat increased due to urban expansion.[52] The 'Dishley Longhorn' resulting from selective breeding became popular toward the late eighteenth century. In this case, historical information is supported by direct archaeozoological evidence: a cache of over 200 cattle horn cores came to light at the site of Cutler Street Warehouses in London dated to between the last quarter of the seventeenth and the late eighteenth century. As the assemblage was dominated by relatively long specimens, it offered a basis for discussing the history of British Longhorn.[53] The popularity of long-horned cattle across Europe may also have been related to an increasing demand for goods such as drinking horns, ladles, and combs made from horn during this relatively late time period. Such objects would have looked far less spectacular when made from the relatively small horn sheets of medieval cattle.

Art and material culture show that, by the eighteenth century, the Hungarian Grey was approaching its modern form. A so-called Habán-style, white glazed faience plate from Hungary (1710) already shows an ox remarkably reminiscent of modern Hungarian Grey: the rather long horns and light colour may be interpreted as representing a form missing from the medieval osteoarchaeological record (Figure 9.7).[54] In 1783, De Bluë published a picture of an ox measuring 6 feet 3 inches tall and 9 feet long showing long horns of 'tulip' conformation.[55] This magnificent beast was exhibited in Einsiedeln, the most important pilgrimage site dedicated to the Virgin Mary in Switzerland. Due to its geographical proximity to the southern German regions

Figure 9.7 White glazed, painted faience plate from Hungary, 1710, photo: Alice M. Choyke.

targeted by Hungarian cattle exports, it is tempting to associate this animal with the trade. However, by this time large cattle had become widely spread in Europe as a result of improved animal breeding techniques. Contemporaneous painted wall tiles from Harlingen (Netherlands, *c.*1780; Museum De Wagg, Deventer) similarly depict a giant ox.[56] The fact that the richly adorned animal is on a similar display shows the appreciation of large, long-horned individuals at the time. In both pictures, the withers heights of the animals are taller than the accompanying humans, giving an artistic emphasis to the importance of size.

Geopolitics and cattle identity

The long-distance cattle trade may have been instrumental in the emergence of the modern Hungarian Grey, but it is only part of the picture, as the breed emerged against the backdrop of the 'General Crisis of the Seventeenth Century'. In the first half of the century, the Thirty Years War (1618–1648) redrew the political map of central Europe and profoundly altered the shape of the Holy Roman Empire. These were also years of Ottoman ascendancy, and Turkish military advances into southeastern Europe occasioned disruption in the livestock trade. In spite of references that Hungarian cattle were preferred to Polish imports in sixteenth-century Vienna, prices varied greatly.[57] The price of beef in general rose in central Europe beginning in the mid-fifteenth century as a consequence of Ottoman occupation of Hungary in the Carpathian Basin.[58] East of the Carpathians, Turkish expansion also affected the Polish livestock trade as it destroyed trade networks in Moldavia, a source of long-distance exports. Meanwhile the establishment of the Union of Lublin (1569) gave Poland access to new territories: in addition to Volhynia and the Kiev region, Podolia was transferred to the Crown of Poland and became a major source of cattle exported toward the west.[59]

Given the resulting intensified competition in the volatile livestock market, horn conformation and distinctive coloration were emphasized as clearly recognizable characteristics of such animals. Such a pronounced association between forms of domestic cattle and the distinct areas from which they had originated was possibly fuelled by inhabitants promoting their trademark produce, perhaps inadvertently exploiting a romantic demand for exotic 'otherness' in economically powerful urban markets. Animal breeds were gradually associated with emerging nations. It would be unrealistic to claim that 'breed' and 'nation' were being used in anything like their modern sense, but these activities clearly demonstrate that they were mutually reinforcing one another during an era in which they began both emerging.

Cattle became a mass medium for constructing and conveying messages of identity. Traded long distance and on a large scale, cattle differed from purebred horses or dogs whose breeding and use were typically elite activities, whose emotional significance was distinctly personalized, and who were often strongly and intimately individuated. Cattle were altogether more collective: a group of animals often represented a breed frequently associated with a region of origin, which was in turn emotionally associated with the groups of humans who formed newly emerging and institutionalized nation states. This relationship, however, did not grow *ex nihilo*. In addition to its regional origins, some cattle gradually became associated with ethnicity through the competitive livestock trade, adding symbolic significance to the high utilitarian value of these animals.

It is worth emphasising, however, that most attributions of 'national character' to domestic animals developed retrospectively. Folk taxonomies distinguishing between regional forms gave place to a systematic inventory of characteristics to be reinforced through targeted selective breeding. Production traits became highlighted by horn conformation and coat colour as definitive trademarks.[60] Before the onset of evolutionary thinking, the resulting image of animals then was taken as the norm understood – albeit erroneously – to be constant. This tendency gained emphasis along the margins of Europe during the nineteenth-century springtime of nation states, in which peoples aspired to form national identities in the face of their competing neighbours asserting ethnic continuities.

From the viewpoint of the buyer, however, 'national' distinctions may have lost significance once the basic qualities of such desirable cattle were consolidated. As Johann Schoen wrote in 1830, 'in [the] Polish cattle trade to Silesia this cattle is called Podolian, the same way as certain wares are known as Nürnberger, although drives from this town are the smallest. In Vienna, the very same cattle is called Hungarian, as they arrive through Hungary'.[61] Likewise, the Brockhaus Lexicon also wrote that 'herds of Polish and Hungarian, actually Podolian cattle are driven to Germany, almost universally of blueish grey colour'.[62] Prior to the First World War, Hungarian Grey breeding stock was already exported to the Balkans where it was used in upgrading local grey cattle.[63] By the late nineteenth century, herd books were established across Western Europe to standardize trademark appearances, partly in response to increasing market competition. Thus, regional and national identities gained market value as represented by strictly defined animal breeds. This may be seen as a reflection of the nineteenth-century consolidation of nation states.

A comparable development, however, followed a hundred years later in Eastern Europe. 'Podolian cattle' became a loosely defined blanket term for a group of breeds, largely characterized by greyish colour and relatively long horns. Aside from Hungarian Grey, Transylvanian, Ukrainian, Moldavian, and Bulgarian Grey cattle, Leopold Adametz described cattle breeds of Podolian character in Serbia and Bosnia.[64] In Italy, Maremman, Romagnola, and Chianina as well as several local breeds (Apulian, Pugliese, Abruzzoso, Montanara, Puglioso del Basso Veneto) are classified within *razza podolica*.[65]

Modern developments

After the First World War, Hungary lost two-thirds of its territory, including peripheral areas where traditional animal keeping was practiced. Figure 9.8 shows the accelerating decline of Hungarian Grey stocks that remained within the new borders in the central Carpathian Basin. By 1925, Hungarian Grey cattle represented only 16.8 percent of the national stock (321,000 head). In 1943, Zoltán Csukás saw the great variability and plasticity of the twentieth-century breed as an asset for modern beef-purpose selection.[66] However, the radical decline continued after the Second World War. The population bottomed out between 1947 and 1967 and approached extinction. The mechanization of agriculture was an inevitable consequence of agricultural collectivization in the Hungarian People's Republic. Draught exploitation, another traditional strength of this breed, became irrelevant and most working Hungarian Grey were slaughtered for beef. During the late 1950s, some 1,800 of the 2,000–3,000 remaining cows were used as breeding material and crossed with sires of the Kostroma dairy breed imported from the

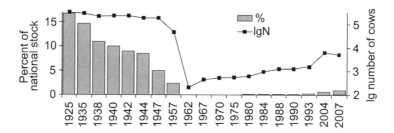

Figure 9.8 Changes in the contribution of Hungarian grey to the national cattle
stocks in Hungary after World War I (author's drawing).

Source: Food and Agriculture Organization of the United Nations (1997), L. Bartosiewicz,
'The Hungarian Grey Cattle: A Traditional European Breed', Animal Genetic Resources
Information. Reproduced with permission.

Soviet Union.[67] According to a 1962 government resolution, only 200 purebred Hungarian Grey cows and six bulls were to be saved.[68] In the face of
this official initiative, there were brave grassroots efforts, often illegal, to save
animals by quietly sabotaging culling or crossing and hiding some breeding
stock at far-away farmsteads. Although the 1962 official numbers would have
been insufficient for maintaining genetic diversity, Hungary subsequently
won international acclaim for its pioneering efforts in the conservation of
ancient breeds.[69]

Hungarian Grey cattle have attracted renewed attention in recent decades.
As with other historic long-horned breeds (among others, Maremman,
English Longhorn, Scottish Highland cattle, and Texas Longhorn), these animals are frequently kept for the kudos attached to their historic image. An up-
surge of nationalism following the political changes of 1989–1991 has visibly
motivated this trend. Recently, this perception attained a new dimension
as free range and lean animals became an important source of high-quality
organic beef. Today the carefully cultivated image of the Hungarian Grey
contributes 'brand value' even to laboratory-approved quality certificates.

Conclusions

In this chapter, the case of Hungarian Grey cattle has been used to illustrate
the historical interactions that have shaped a distinctive high-status animal
and *topoi* regarding its development as a national symbol. Behind this particular example lies a multi-faceted historical process whereby animal and human
identities were mutually constitutive. Breed and nation are analogous modern
concepts that have often been applied anachronistically, yet human and animal identities influenced one another while these categories were emerging
and long before they became more or less fixed in their modern meanings. As
argued here, the persistent attention paid to Hungarian Grey cattle during the
last two centuries and the emotional commitment to the millennium-long

continuity of this breed are prime instances of this process of the mutual constitution of identities at work. The other widely held belief is that the Hungarian Grey cattle is direct heir to the now-extinct aurochs. Although neither of these hypotheses has ever been supported by osteological evidence from archaeological sites, they both stand strong in national mythology.

While natural evolution is random, humans have imposed both conscious and subconscious changes on animals ever since the first species were domesticated. This process is perpetual and irreversible. In contrast to popular belief, Hungarian Grey cattle must have emerged relatively recently as a distinct breed and may be considered a spin-off of the late medieval and early modern (largely Ottoman) cattle trade. Market forces would have stimulated selective breeding for its imposing appearance, especially the long horns and silver colour of the coat, a kind of early modern 'branding'. Intensive movement of highly variable late medieval stocks provided an excellent basis for selection in these directions. Market driven, targeted breeding possibly played a decisive role in consolidating this trademark appearance, which became common by the eighteenth century.

As the 'product' was perfected, it exerted a continuous feedback on the creation process. Of course, mutualism between breeders and animals was asymmetric, overshadowed by raw financial interest. However, the emergence of national consciousness in the latter part of the early modern era provided a new ideological context in which Hungarian Grey cattle attained increasing significance as the embodiment of a particular national identity. The putative autochthony of the Hungarian Grey reinforced and was reinforced by Hungarian nationalism, in a process that is still ongoing.

Notes

1   Charles R. Darwin, *The Variation of Animals and Plants Under Domestication*, 2 vols (London: John Murray, 1868), i, 80.

2   László Bartosiewicz, Wim Van Neer, and An Lentacker, *Draught Cattle: Their Osteological Identification and History* (Annalen. Zoologische Wetenschappen, 281; Tervuren: Koninklijk Museum voor Midden-Afrika, 1997).

3   Tunç Sipahi, 'New Evidence From Anatolia Regarding Bull Leaping Scenes in the Art of the Aegean and the Near East', *Anatolica*, 27 (2001), 107–25.

4   Frederick Monderson, *Hatshepsut's Temple at Deir El Bahari* (Bloomington: AuthorHouse, 2007).

5   *Iliad,* 14.155.

6   Also called sometimes Hungarian White in English. See, for example, Sándor Bökönyi, *History of Domestic Animals in Central and Eastern Europe* (Budapest: Akadémiai Kiadó, 1974), 139.

7   Imre Bodó, István Gera and Gábor Koppány, *A magyar szürke marha* [The Hungarian Grey Cattle] (Budapest: A Magyar Szürke Szarvasmarhát Tenyésztők Egyesülete, 1994), 22.

8   László Bartosiewicz, 'The Hungarian Grey Cattle: A Traditional European Breed', *Animal Genetic Resources Information*, 21 (1997), 51.

9   Ottó Herman, *A magyar pásztorok nyelvkincse* [The Vocabulary of Hungarian herdsmen] (Budapest: Hornyánszky, 1914), 453–61.

10 Béla Tormay, *A szarvasmarha és tenyésztése I–II* [Cattle breeding] (Budapest: Athenaeum Irodalmi és Nyomdai R. T., 1901), 42.

11 Imre Bodó (ed.), *Horn Conformations* (Budapest: Department of Animal Husbandry, University of Veterinary Science, 1991).

12 Bodó, Gera and Koppány, *A magyar szürke marha*, 35.

13 László Bartosiewicz, 'Régi kor árnya felé visszamerengni mit ér? A skót hegyi marha és a romantika [Is it worth dreaming about the past? The Scottish Highland cattle and romanticism]', in Levente Jávorka (ed.), *"Emberségről példát"*. *Válogatott közlemények Bodó Imre 80. születésnapjára* (Budapest: a Magyar Szürke Marhát Tenyésztők Egyesületének és a Magyar Szürke Marhát Kedvelők Baráti Körének Kiadványa, 2012), 39.

14 Other breeds involved in Heinz's scheme were Angeln cattle, Black-pied lowland cattle, Brown Swiss, Corsican cattle, Scottish Highland cattle, Murnau-Werdenfels cattle, and White Park (Chillingham) cattle. Lutz Heck, 'Über die Neuzüchtung des Ur oder Auerochs', *Berichte, Internationalen Gesellschaft zur Erhaltung des Wisents*, 3/4 (1934), 225–94; Heinz Heck, 'Über den Auerochsen und seine Rückzüchtung', *Jahrbuch des Vereins für Naturkunde im Herzogtum Nassau*, 90 (1952), 107–24.

15 László Bartosiewicz, Alexandra Gyetvai and Hans-Christian Küchelmann, 'Beast in the Feast', in Aleksander Pluskowski et al. (eds), *Bestial Mirrors: Animals as Material Culture in the Middle Ages* (ViaVIAS, 3; Vienna: VIAS, 2010), 92.

16 Lutz Heck, 'Der Ur', *Das Tier und Wir, Monatszeitschrift für alle Tierfreunde*, 5/3 (Mar. 1934), 1–16.

17 Jon Kay, '"Nazi" Cattle: Farmer Forced to Cull Part of Herd', BBC News (online), 6 Jan. 2015, www.bbc.com/news/uk-30703082, accessed 14 Jul. 2015.

18 Simon de Bruxelles, 'A Shaggy Cow Story: How A Nazi Experiment Brought Extinct Aurochs to Devon', The Times (online), 22 Apr. 2009, www.thetimes.co.uk/tto/news/uk/article1967283.ece, accessed 29 Sep. 2016.

19 Boria Sax, *Animals in the Third Reich: Pets, Scapegoats, and the Holocaust* (Providence: Yogh & Thorn, 2013), 83.

20 Ludwig Rütimeyer, 'Über Art und Rasse des zahmen europäischen Rindes', *Archiv für Anthropologie*, 1 (1866), 219–50.

21 Darwin, *Variation of Animals*, 80.

22 Miklós Jankovich, 'Adatok a magyar szarvasmarha eredetének és hasznosításának kérdéséhez', *Agrártörténeti Szemle*, 3/4 (1967), 420–31.

23 The concept of nation, known to have been a bourgeois development, would have been ideologically far more contentious to promote by the time of the communist regime.

24 István Vörös, 'Early Medieval Aurochs (*Bos Primigenius* Boj.) and His Extinction in Hungary', *Folia Archaeologica*, 34 (1985), 193–218; László Bartosiewicz, 'Hungarian Grey cattle: in search of origins', *Hungarian Agricultural Research*, 5/3 (1996), 4–9.

25 László Bartosiewicz, 'The Emergence of Holocene Faunas in the Carpathian Basin: A Review', in Norbert Benecke (ed.), *The Holocene History of the European Vertebrate Fauna* (Archäologie in Eurasien; 6; Rahden: Verlag Marie Leidorf GmbH, 1999), 80.

26 János Matolcsi, 'Historische Erforschung der Körpergrösse des Rindes auf Grund von ungarischem Knochenmaterial', *Zeitschrift für Tierzüchtung und Züchtungsbiologie*, 63 (1970), 155–94; Bökönyi, *History*.

27 Ceiridwen J. Edwards et al., 'Mitochondrial DNA Shows a Near Eastern Neolithic Origin of Domestic Cattle and No Indication of Domestication of European Aurochs', *Proceedings of the Royal Society B.*, 274 (2007), 1377–785.

28 László Bartosiewicz, 'A magyar szürke marha históriája [History of the Hungarian grey cattle]', *Természet Világa*, 124/2 (1993), 54–7. The possibility that

marauding mounted warriors brought back such animals has rarely been considered. See Géza Ferencz, 'Őshonos, ősi magyar vagy ősi jellegű állatunk-e a magyar szürke marha? [Is Hungarian grey an autochtonous, ancient or ancient-like cattle breed?]', *Állattenyésztés*, 25/4 (1976), 363–78.

29 In this popular narrative, the term autochthonous is erroneously used as it means animals originating in the place where found. See László Bartosiewicz, 'Are "Autochthonous" Animal Breeds Living Monuments?', in Erzsébet Jerem–Zsolt Mester and Réka Benczes (eds), *Archaeological and Cultural Heritage Preservation within the Light of New Technologies* (Budapest: Archaeolingua, 2006), 33.

30 László Bartosiewicz, Dóra Mérai and Péter Csippán, 'Dig up–Dig in: Practice and Theory in Hungarian Archaeology', in Ludomir R. Lozny (ed.), *Comparative Archaeologies: A Sociological View of the Science of the Past* (New York: Springer, 2011), 309.

31 László Bartosiewicz, 'Huns', in Peter Bogucki and Pam J. Crabtree (eds), *Ancient Europe, 8000 B.C. to A.D. 1000: An Encyclopedia of the Barbarian World* (New York: Scribners' Sons, 2004), 391–93. For many, conflating Hunnic and ancient Hungarian identities under the label of Attila, the so-called '*flagellum dei*', has been an integral part of national stereotyping since the nineteenth century.

32 János Matolcsi, *A háziállatok eredete* [The origins of domestic animals] (Budapest: Mezőgazdasági Kiadó, 1975); László Bartosiewicz, 'A Millennium of Migrations: Protohistoric Mobile Pastoralism in Hungary', *Bulletin of the Florida Museum of Natural History*, 44 (2003), 33.

33 Éva Ágnes Nyerges, 'Ethnic Traditions in Meat Consumption and Herding at a 16th Century Cumanian Settlement in the Great Hungarian Plain', in Sharyn Jones O'Day, Wim van Neer and Anton Ervynck (eds), *Behaviour Behind Bones: The Zooarchaeology of Ritual, Religion, Status and Identity* (Oxford: Oxbow 2004), 262–70.

34 Bartosiewicz, Mérai and Csippán, 'Dig up–Dig in', 284.

35 Zoltán Csukás, *A podoliai marhacsoport az Apennini félszigeten* [The Podolian group of cattle on the Apennine Peninsula] (Debrecen: Pallag, 1943).

36 Bökönyi, *History*, 142.

37 Imre Bodó, István Gera and Gábor Koppány, *The Hungarian Grey cattle breed* (Budapest: Association of the The Hungarian Grey Cattle Breeders, 1996).

38 Cattle from Scania and Denmark ensured the meat supply of urban centres along the Rhine, in Brabant, and Flanders, and similar movements were documented in Britain. See Ian Blanchard, 'The Continental European Cattle Trades 1400–1600', *Economic History Review*, 2nd ser., 39 (1986), 445; László Bartosiewicz, 'Régi kor árnya felé visszamerengni mit ér?', 28.

39 Sándor Takáts, *Szegény magyarok* [Poor Hungarians] (Budapest: Genius, 1927); Alajos Miskulin, *Magyar művelődéstörténeti mozzanatok Giovanni és Mateo Villani krónikái alapján* [Moments in Hungarian cultural history on the basis of chronicles by Giovanni and Mateo Villani] (Budapest: Stephaneum, 1905).

40 Sándor Milhoffer, *Magyarország közgazdasága I* [The Economy of Hungary I] (Budapest: Franklin, 1904); László Bartosiewicz, *Animals in the Urban Landscape in the Wake of the Middle Ages* (Oxford: BAR International Series 609, 1995), Table 32.

41 Bartosiewicz, *Animals*, 84, Table 33.

42 In early modern Scottish myth, Rob Roy embodies a similar stereotype. See Bartosiewicz, 'Régi kor árnya felé visszamerengni mit ér?', 36.

43 László Bartosiewicz, 'Cattle Trade across the Danube at Vác (Hungary)', *Anthropozoologica*, 21 (1995), 189–95. Southwest-bound herds from Hungary advanced toward Croatia along the Drava River or crossed the Adriatic Sea by boat between Zadar and Venice. László Gaál, *A magyar állattenyésztés múltja* [The past of Hungarian animal breeding] (Budapest: Mezőgazdasági Kiadó, 1966), 475; László Bartosiewicz, 'Turkish Period Bone Finds and Cattle Trade in South-Western

Hungary', in Cornelia Becker et al. (eds), *Historia animalum ex ossibus* (Beiträge zur Paläoanatomie, Archäologie (Ägyptologie, Ethnologie und Geschichte der Tiermedizin; Rahden: Verlag Marie Leidorf GmbH, 1999), 48, Fig. 1.

44 Blanchard, 'Continental European Cattle Trades', 433.

45 Due to the vicissitudes of Hungary's history, the Hungarian Grey cannot be compared to the archetype of all autochthonous cattle, the Chillingham breed, thought to have been formed over at least 700 years within the same, undisturbed estate. Lawrence Alderson, *The Chance to Survive* (Northampton: A. H. Jolly, 1989).

46 Bökönyi, *History*, 442.

47 Péter Csippán, '18. századi szarvcsapleletek a budai Vízivárosból' [Eighteenth-century cattle horn core finds from the Víziváros District of Buda, Hungary], in László Bartosiewicz, Erika Gál and István Kováts (eds), *Csontvázak a szekrényből. Válogatott tanulmányok a Magyar Archaeozoológusok Visegrádi Találkozóinak anyagából 2002–2009* (Budapest: Martin Opitz Kiadó 2009), 163, Fig. 5.

48 Péter Csippán, '18. századi szarvcsapleletek a budai Vízivárosból', 167, Fig. 5.

49 Matolcsi, 'Historische Erforschung'; Bökönyi, *History*; István Vörös, 'Egy 15. századi ház csontlelete Vácott' [Animal bones from a 15th century house in Vác], *Archeologiai Értesítő*, 113 (1986), 256.

50 Mihály Petri, *Szilágy megye topographiája* [The topography of Szilágy County] (Budapest: Szilágy Vármegye Közönsége, 1904); Ottó Herman, *A magyarok nagy ősfoglalkozása* [The great ancient trade of Hungarians] (Budapest: Hornyánszky, 1909).

51 Bartosiewicz, *Animals*, 48, Fig. 27.

52 Ian L. Mason, *A World Dictionary of Livestock Breeds, Types and Varieties* (4th edn, Wallingford: C. A. B. International, 1996).

53 Philip L. Armitage, 'A System for Ageing and Sexing the Horn Cores of Cattle from British Postmedieval Sites (with Special Reference to Unimproved British Longhorn Cattle)', in Bob Wilson, Caroline Grigson and Sebastian Payne (eds), *Ageing and sexing animal bones from archaeological sites* (BAR, British Series, 109; Oxford: British Archaeological Reports, 1982), 37–54.

54 Note that this piece of art pre-dates by only one year what is considered the 'end' of archaeological periods in Hungary.

55 Vera De Bluë, *Landaus-landab* (Bern–Munich: Edition Erpf, 1985), 65.

56 Johan ten Broeke, Arno van Sabben and Richard Woudenberg, *Antieke tegels–Antique tiles–Carreaux anciens* (Oud-Beijerland: Van As, 2000), 138.

57 Sándor Takáts, *Emlékezzünk eleinkről* [Remembering our ancestors] (Budapest: Genius, 1929), 334.

58 Slicher van Bath, Bernard Hendrik, *The Agrarian History of Western Europe A. D. 1500–1850* (London: Edward Arnold, 1963), 204; Friedrich Lütge, *Strukturwandlungen im ostdeutschen und osteuropäischen Fernhandel des 14. bis 16. Jahrhunderts* (Munich: Verlag der Bayerischen Akademie der Wissenschaften, 1964), 40.

59 Podolia is stretched south of Volhynia, southwest of the Kiev Region and northeast of the Dniester River. On the role of Podolian cattle exports see Blanchard, 'Continental European Cattle Trades', 440.

60 The final consolidation of physical traits in Hungarian Grey and Scottish Highland cattle shows parallels and differences related to the situations under which these breeds evolved during the eighteenth to twentieth centuries. Bartosiewicz, 'Régi kor árnya felé visszamerengni mit ér?'.

61 Johann Schoen, *Staatswirthschaftliche Berechnungen in Bezug auf die Vieh-Zölle und Quarantäne Preussens insbesondre Schlesiens* (Breslau, 1830), 49. All translations are the author's own.

62 Friedrich Arnold Brockhaus, *Bilder-Conversations-Lexikon, Ein Handbuch zur Verbreitung gemeinnütziger Kenntnisse und zur Unterhaltung. Erster Band* (Leipzig, 1837), 713–15.

63 János Mattesz, *A mezőhegyesi magyarfajta marha monográfiája* [Monograph Study of the Hungarian Cattle in Mezőhegyes] (Sopron: Székely és Társa Könyvnyomdája, 1927).

64 Leopold Adametz, *Die Rinderrassen und Schläge in Bosnien, der Herzegovina und im nördlichen Theile des Sandschaks von Novibazar* (Vienna: Sonderdruck, 1892).

65 Bodó, Gera and Koppány, *The Hungarian Grey Cattle Breed*, 33.

66 Csukás, *A podoliai marhacsoport*.

67 József Schandl, *Szarvasmarhatenyésztés* (Budapest: Mezőgazdasági Kiadó, 1962).

68 Bodó, Gera and Koppány, *The Hungarian Grey Cattle Breed*.

69 Alderson, *The Chance to Survive*; Hans-Peter Grünenfelder, 'Protection of Genetic Resources in Eastern Europe', *American Livestock Breeds Conservancy News*, 11 (1994), 16–17.

Bibliography

Adametz, Leopold, *Die Rinderrassen und Schläge in Bosnien, der Herzegovina und im nördlichen Theile des Sandschaks von Novibazar* (Vienna: Sonderdruck, 1892).

Alderson, Lawrence, *The Chance to Survive* (Northampton: A. H. Jolly, 1989).

Armitage, Philip L., 'A System for Ageing and Sexing the Horn Cores of Cattle from British Postmedieval Sites (with Special Reference to Unimproved British Longhorn Cattle)', in Bob Wilson, Caroline Grigson and Sebastian Payne (eds), *Ageing and Sexing Animal Bones* (Oxford: British Archaeological Reports, 1982), 37–54.

Bartosiewicz, László, 'A magyar szürke marha históriája [History of the Hungarian Grey Cattle]', *Természet Világa*, 124/2 (1993), 54–57.

———, *Animals in the Urban Landscape in the Wake of the middle ages* (Oxford: BAR International Series 609, 1995).

———, 'Cattle Trade across the Danube at Vác (Hungary)', *Anthropozoologica*, 21 (1995), 189–96.

———, 'Hungarian Grey Cattle: In Search of Origins', *Hungarian Agricultural Research*, 5/3 (1996), 4–9.

———, 'The Hungarian Grey Cattle: A Traditional European Breed', *Animal Genetic Resources Information*, 21 (1997), 49–60.

———, Wim Van Neer and An Lentacker, *Draught Cattle: Their Osteological Identification and History* (Annalen. Zoologische Wetenschappen, 281; Tervuren: Koninklijk Museum voor Midden-Afrika, 1997).

———, 'The Emergence of Holocene Faunas in the Carpathian Basin: A Review', in Norbert Benecke (ed.), *Holocene History of the European Vertebrate Fauna* (Rahden: Verlag Marie Leidorf GmbH, 1999), 73–90.

———, 'Turkish Period Bone Finds and Cattle Trade in South-Western Hungary', in Cornelia Becker et al. (eds), *Historia animalum ex ossibus* (Rahden: Verlag Marie Leidorf GmbH, 1999), 47–56.

———, 'A Millennium of Migrations: Protohistoric Mobile Pastoralism in Hungary', *Bulletin of the Florida Museum of Natural History*, 44 (2003), 101–30.

———, 'Huns', in Peter Bogucki and Pam J. Crabtree (eds), *Ancient Europe, 8000 B.C. to A.D. 1000: An Encyclopedia of the Barbarian World* (New York: Scribners' Sons, 2004), 391–93.

———, 'Are "Autochthonous" Animal Breeds Living Monuments?', in Erzsébet Jerem–Zsolt Mester and Réka Benczes (eds), *Archaeological and Cultural Heritage Preservation within the Light of New Technologies* (Budapest: Archaeolingua, 2006), 33–47.

————, Erika Gál and István Kováts (eds), *Csontvázak a szekrényből. Válogatott tanulmányok a Magyar Archaeozoológusok Visegrádi Találkozóinak anyagából 2002–2009* (Budapest: Martin Opitz Kiadó, 2009).

————, Alexandra Gyetvai and Hans-Christian Küchelmann, 'Beast in the Feast', in Aleksander Pluskowski et al. (eds), *Bestial Mirrors: Animals as Material Culture in the Middle Ages* (Vienna: VIAS, 2010), 85–99.

————, Dóra Mérai and Péter Csippán, 'Dig up–Dig in: Practice and Theory in Hungarian Archaeology', in Ludomir R. Lozny (ed.), *Comparative Archaeologies: A Sociological View of the Science of the Past* (New York: Springer, 2011), 273–337.

————, 'Régi kor árnya felé visszamerengni mit ér? A skót hegyi marha és a romantika [Is It Worth Dreaming about the Past? The Scottish Highland Cattle and Romanticism]', in Levente Jávorka (ed.), *Emberségről példát* (Budapest: a Magyar Szürke Marhát Tenyésztők Egyesületének és a Magyar Szürke Marhát Kedvelők Baráti Körének Kiadványa, 2012), 24–47.

Becker, Cornelia et al. (eds), *Historia animalum ex ossibus* (Beiträge zur Paläoanatomie, Archäologie, Ägyptologie, Ethnologie und Geschichte der Tiermedizin; Rahden: Verlag Marie Leidorf GmbH, 1999).

Benecke, Norbert (ed.), *The Holocene History of the European Vertebrate Fauna* (Archäologie in Eurasien, 6; Rahden: Verlag Marie Leidorf GmbH, 1999).

Blanchard, Ian, 'The Continental European Cattle Trades 1400–1600', *Economic History Review*, 2nd ser., 39 (1986), 427–60.

Bodó, Imre (ed.), *Horn Conformations* (Budapest: Department of Animal Husbandry, University of Veterinary Science, 1991).

Bodó, Imre, István Gera and Gábor Koppány, *A magyar szürke marha* [The Hungarian Grey Cattle] (Budapest: A Magyar Szürke Szarvasmarhát Tenyésztők Egyesülete, 1994).

————, *The Hungarian Grey Cattle Breed* (Budapest: Association of the Hungarian Grey Cattle Breeders, 1996).

Bogucki, Peter and Pam J. Crabtree (eds), *Ancient Europe, 8000 B.C. to A.D. 1000: An Encyclopedia of the Barbarian World* (New York: Scribners' Sons, 2004).

Bökönyi, Sándor, *History of Domestic Animals in Central and Eastern Europe* (Budapest: Akadémiai Kiadó, 1974).

Brockhaus, Friedrich Arnold, *Bilder-Conversations-Lexikon, Ein Handbuch zur Verbreitung gemeinnütziger Kenntnisse und zur Unterhaltung. Erster Band* (Leipzig, 1837).

Bruxelles, Simon de, 'A Shaggy Cow Story: How a Nazi Experiment Brought Extinct Aurochs to Devon', *The Times*, 22 Apr. 2009, www.thetimes.co.uk/tto/news/uk/article1967283.ece, accessed 29 Sep. 2016.

Csippán, Péter, '18. századi szarvcsapleletek a budai Vízivárosból (Eighteenth-Century Cattle Horn Core Finds from the Víziváros District of Buda, Hungary)', in László Bartosiewicz, Erika Gál and István Kováts (eds), *Csontvázak a szekrényből. Válogatott tanulmányok a Magyar Archaeozoológusok Visegrádi Találkozóinak anyagából 2002–2009* (Budapest: Martin Opitz Kiadó 2009), 163–69.

Csukás, Zoltán, *A podoliai marhacsoport az Apennini félszigeten* [The Podolian Group of Cattle on the Apennine Peninsula] (Debrecen: Pallag, 1943).

Darwin, Charles R., *The Variation of Animals and Plants Under Domestication*, 2 vols (London: John Murray, 1868).

De Blue, Vera, *Landaus-landab* (Bern–München: Edition Erpf, 1985).

Edwards, Ceiridwen J. et al., 'Mitochondrial DNA Shows a Near Eastern Neolithic Origin of Domestic Cattle and No Indication of Domestication of European Aurochs', *Proceedings of the Royal Society B.*, 274 (2007), 1377–785.

Ferencz, Géza, 'Őshonos, ősi magyar vagy ősi jellegű állatunk-e a magyar szürke marha? [Is Hungarian Grey an Autochtonous, Ancient or Ancient-Like Cattle Breed?]' *Állattenyésztés*, 25/4 (1976), 363–78.

Gaál, László, *A magyar állattenyésztés múltja* [The Past of Hungarian Animal Breeding]. (Budapest: Mezőgazdasági Kiadó, 1966).

Grünenfelder, Hans-Peter, 'Protection of Genetic Resources in Eastern Europe', *American Livestock Breeds Conservancy News*, 11 (1994), 16–17.

Heck, Heinz, 'Über den Auerochsen und seine Rückzüchtung', *Jahrbuch des Vereins für Naturkunde im Herzogtum Nassau*, 90 (1952), 107–24.

Heck, Lutz, 'Der Ur', *Das Tier und Wir, Monatszeitschrift für alle Tierfreunde*, 5/3 (Mar. 1934), 1–16.

———, 'Über die Neuzüchtung des Ur oder Auerochs', *Berichte, Internationalen Gesellschaft zur Erhaltung des Wisents*, 3/4 (1934), 225–94.

Herman, Ottó, *A magyarok nagy ősfoglalkozása* [The Great Ancient Trade of Hungarians] (Budapest: Hornyánszky, 1909).

———, *A magyar pásztorok nyelvkincse* [The Vocabulary of Hungarian Herdsmen] (Budapest: Hornyánszky, 1914).

Jankovich, Miklós, 'Adatok a magyar szarvasmarha eredetének és hasznosításának kérdéséhez', *Agrártörténeti Szemle*, 3/4 (1967), 420–31.

Jávorka, Levente (ed.), *"Emberségről példát". Válogatott közlemények Bodó Imre 80. születésnapjára* (Budapest: a Magyar Szürke Marhát Tenyésztők Egyesületének és a Magyar Szürke Marhát Kedvelők Baráti Körének Kiadványa, 2012).

Jones O'Day, Sharyn, Wim van Neer and Anton Ervynck (eds), *Behaviour Behind Bones: The Zooarchaeology of Ritual, Religion, Status and Identity* (Oxford: Oxbow 2004).

Kay, Jon, '"Nazi" Cattle: Farmer Forced to Cull Part of Herd', BBC News, 6 Jan. 2015, www.bbc.com/news/uk-30703082, accessed 14 Jul. 2015.

Lozny, Ludomir R. (ed.), *Comparative Archaeologies: A Sociological View of the Science of the Past* (New York: Springer, 2011).

Lütge, Friedrich, *Strukturwandlungen im ostdeutschen und osteuropäischen Fernhandel des 14. bis 16. Jahrhunderts* (Munich: Verlag der Bayerischen Akademie der Wissenschaften, 1964).

Mason, Ian L., *A World Dictionary of Livestock Breeds, Types and Varieties* (4th edn, Wallingford: C. A. B. International, 1996).

Matolcsi, János, 'Historische Erforschung der Körpergrösse des Rindes auf Grund von ungarischem Knochenmaterial', *Zeitschrift für Tierzüchtung und Züchtungsbiologie*, 63 (1970), 155–94.

———, *A háziállatok eredete* [The Origins of Domestic Animals] (Budapest: Mezőgazdasági Kiadó, 1975).

Mattesz, János, *A mezőhegyesi magyarfajta marha monográfiája* [Monograph Study of the Hungarian Cattle in Mezőhegyes] (Sopron: Székely és Társa Könyvnyomdája, 1927).

Mester, Erzsébet Jerem–Zsolt and Réka Benczes (eds), *Archaeological and Cultural Heritage Preservation within the Light of New Technologies* (Budapest: Archaeolingua, 2006).

Milhoffer, Sándor, *Magyarország közgazdasága I* [The Economy of Hungary I] (Budapest: Franklin, 1904).

Miskulin, Alajos, *Magyar művelődéstörténeti mozzanatok Giovanni és Mateo Villani krónikái alapján* [Moments in Hungarian Cultural History on the Basis of Chronicles by Giovanni and Mateo Villani] (Budapest: Stephaneum, 1905).

Monderson, Frederick, *Hatshepsut's Temple at Deir El Bahari* (Bloomington IN: AuthorHouse, 2007).

Nyerges, Éva Ágnes, 'Ethnic Traditions in Meat Consumption and Herding at a 16th Century Cumanian Settlement in the Great Hungarian Plain', in Sharyn Jones O'Day, Wim van Neer and Anton Ervynck (eds), *Behaviour Behind Bones: The Zooarchaeology of Ritual, Religion, Status and Identity* (Oxford: Oxbow 2004), 262–70.

Petri, Mihály, *Szilágy megye topographiája* [The Topography of Szilágy County] (Budapest: Szilágy Vármegye Közönsége, 1904).

Pluskowski, Aleksander et al. (eds), *Bestial Mirrors: Animals as Material Culture in the Middle Ages* (ViaVIAS, 3; Vienna: VIAS, 2010).

Rütimeyer, Ludwig, 'Über Art und Rasse des zahmen europäischen Rindes', *Archiv für Anthropologie*, 1 (1866), 219–50.

Sax, Boria, *Animals in the Third Reich: Pets, Scapegoats, and the Holocaust* (Providence: Yogh & Thorn, 2013).

Schandl, József, *Szarvasmarhatenyésztés* (Budapest: Mezőgazdasági Kiadó, 1962).

Schoen, Johann, *Staatswirthschaftliche Berechnungen in Bezug auf die Vieh-Zölle und Quarantäne Preussens insbesondre Schlesiens* (Breslau, 1830).

Sipahi, Tunç, 'New Evidence From Anatolia Regarding Bull Leaping Scenes in the Art of the Aegean and the Near East', *Anatolica*, 27 (2001), 107–25.

Slicher van Bath, Bernard Hendrik, *The Agrarian History of Western Europe A. D. 1500–1850* (London: Edward Arnold, 1963).

Takáts, Sándor, *Szegény magyarok* [Poor Hungarians] (Budapest: Genius, 1927).

———, *Emlékezzünk eleinkről* [Remembering Our Ancestors] (Budapest: Genius 1929).

ten Broeke, Johan, Arno van Sabben and Richard, Woudenberg, *Antieke tegels–Antique tiles–Carreaux anciens* (Oud-Beijerland: Van As, 2000).

Tormay, Béla, *A szarvasmarha és tenyésztése I–II* [Cattle breeding] (Budapest: Athenaeum Irodalmi és Nyomdai R. T., 1901).

Vörös, István, 'Early Medieval Aurochs (*Bos primigenius* Boj.) and His Extinction in Hungary', *Folia Archaeologica*, 34 (1985), 193–218.

———, 'Egy 15. századi ház csontlelete Vácott [Animal bones from a 15th century house in Vác]', *Archeologiai Értesítő*, 113 (1986), 255–56.

Wilson, Bob, Caroline Grigson and Sebastian Payne (eds), *Ageing and Sexing Animal Bones from Archaeological Sites* (BAR, British Series, 109; Oxford: British Archaeological Reports, 1982).

10 'The Monster's Mouth...'

Dangerous animals and the European settlement of Australia

Krista Maglen

In 1803, David Collins was put in command of establishing a new penal colony on the southern coast of Australia. Aimed primarily at averting French claims to the south, on 9 October Collins settled over 400 people, including marines, free settlers and nearly 300, mostly male, prisoners, on the Nepean Peninsula in Port Philip Bay. From the beginning, the 'canvas town' encampment of Sullivan Bay struggled in the sandy, scrappy landscape that provided little access to fresh water and few animals to hunt.[1] Within only a few weeks Collins could see that without decisive action, the colony was likely to tear itself apart as the marines grew mutinous and the convicts looked for routes of escape. Asserting his control, Collins had an insubordinate soldier flogged in front of the entire colony with 700 lashes.[2] But with only 50 unhappy marines to guard the increasingly rebellious convicts, eight of whom had attempted escape within the first month, managing the rest of the population was a more difficult task.[3] Without the all-seeing and impenetrable Panopticon that Collins had been encouraged to construct in the numerous letters he had received from Jeremy Bentham before he left England, the calm seas of the bay and gentle slopes of the peninsula seemed to offer little obstacle to fleeing felons. But, to a commander with twelve years' experience in New South Wales, where he had been Judge Advocate and Lieutenant Governor from 1788 to 1796, the environment offered other custodial defences. On 20 October, Collins issued a General Order stating that

> This Bay and the Harbour in general – being unfortunately full of voracious Sharks and Stingrays only – it is recommended to the convicts not to go into the water without the utmost precaution and they are positively prohibited from bathing in front of the Encampment.[4]

The danger of sharks and other marine creatures was well known and understood by Europeans, and the presence of sharks in Sydney Harbour would have been noted from the earliest days of settlement. Indeed, Watkin Tench, a marine and chronicler of the settlement at Sydney Cove, described in 1791 the orphaning of an Aboriginal boy whose mother had been 'bitten in two by a shark'.[5] However, reports of early shark attacks on people are scarce (although at least one dog fell victim during the first decade). Yet, in both Port Phillip Bay and in Sydney, as well as in the other convict settlements, the 'monsters' that lurked beneath the surface of the surrounding seas were quickly understood as threatening gatekeepers to the coastal edges of the colonies. The perceived risk posed by sharks rather than the reality of attacks was what defined the first European encounters.

This chapter explores the relationships that Europeans in Australia established with dangerous creatures like sharks and snakes during the first century of settlement and considers the role of these animals in contributing to European framings and understandings of the natural and built environment. It also considers what thinking about animals in this way does to their place in the historical narrative. How, for example, might we think about David Collins's enlisting of the sharks and stingrays of Port Philip Bay: as a rational warning against 'natural' hazards, or as a call to actors to participate in the policing of the penal outpost's collapsing order? Can we think of the sharks of Port Philip Bay as part of the history of the colony, or because there were no 'attacks' during the failed occupation of Sullivan Bay, are they merely incidental to the story?

The scale of the danger presented by marine creatures was deliberately overstated as a means of deterring convict escape in Port Philip as it was in other penal settlements. Fear of sharks was well established by the late eighteenth century, and Europeans had learned of their potential ferocity since travels into the American tropics had brought them into contact with some of the larger species of sharks, unknown in European waters.[6] With many sailors among both the free and convict population, first-hand or community knowledge of shark behaviour would have existed within the early settlements. Indeed, as James Tuckey, who accompanied David Collins during the Sullivan Bay expedition, stated in his account, 'the shark is the hereditary foe of sailors; and the moment one is spied, the whole crew are instantly in arms'.[7]

Superimposed onto these fears were the warnings of penal and colonial officials, who were keen to utilise these creatures as guards around the periphery of the colonies, as we have just seen in the Sullivan Bay settlement. We can see the same recruiting of sharks as sentries along Eaglehawk Neck, the narrow land bridge – a mere 30 meters across – that linked the penal station at Port Arthur in Van Diemen's Land (later to be renamed as Tasmania) to the mainland. Guarding against escape attempts, soldiers were stationed behind up to eighteen watchdogs, driven to savagery through the appalling conditions in which they were kept, entrusted with warning or attacking any runaways.[8] In addition, three platforms were constructed off the beach where more dogs raised the alert to convicts who attempted to avoid detection by wading through the sea.[9] Those fugitives who did manage to evade Eaglehawk Neck's dogs and lamps and marines, it was said, 'would have to battle the watch with an outlying picquet of sharks abounding in these waters'.[10] The notorious escapee William Westwood, often mistaken in the records for an Aboriginal man because of his nickname 'Jackey Jackey', survived the sharks of the Neck in his breakout from Port Arthur in 1842, but his three companions were not so fortunate. Numerous sources cite the incident, one from 1852 describing how 'four absconders faced the rolling surf (three English, one a negro); the white swimmers were seized by a no less formidable guardian of the waters than a rapacious shark. The darkey [Westwood] got safe to land, but was taken by the outlying piquet'.[11] Stories such as this were well known and actively disseminated, adding to official warnings and established fears about shark-infested waters. But, in order to ensure the participation of the predatory fish in the cold waters of southern Tasmania, offal from the nearby slaughterhouse was thrown into the water in order to attract them to their sentry posts along the Neck.[12]

So, the sharks joined the marines and dogs in regulating the penal coastline primarily through the threat they represented, rather than through the more exceptional occurrence that befell Westwood's companions. Drawing on fears of a well-known adversary and exaggerating them proved to be a useful tool in the policing of these generally understaffed imperial outposts. Snakes, on the other hand, which caused many more deaths among early settlers than sharks, were not as frequently deployed. This may be accounted for in a number of ways.

Because the British had historical and imperial experience with snakes they knew that, unlike large sharks, not all species posed potential danger to human life. And so, the very earliest accounts noted the presence of snakes but questioned whether they were venomous. Unlike sharks, their potency was not assumed. First Fleeter John White, in the 1790 publication of his journal, declared that, 'none of the snakes appear to be of a poisonous nature'.[13] While Tench remarked in his account that

> of snakes there are two or three sorts: but whether the bite of any of them be mortal, or even venomous, is somewhat doubtful. I know but of one well attested instance of a bite being received from a snake. A soldier was bitten so as to draw blood, and the wound healed as a simple incision usually does without shewing any symptom of malignity.[14]

He went on to note the quick and painful death by snakebite of another abject colonial dog, and the 'utmost horror' expressed among the Sydney Aboriginal community at the sight of a snake but, regardless, concluded that they were not much to worry about. However, it was not long before the venomous effect of native serpents was made clear.

There are no records to be found of a fatal bite until 1792 when, in January of that year, at the height of the summer, a convict fainted while pulling a cart of bricks. David Collins, writing over a decade before his command at Sullivan Bay, describes how the man was laid down in the shade by the side of the road where, in his exhaustion, he fell asleep. Sometime later, he awoke to feel something wrapped tightly around his neck, and putting his hand to his throat he discovered that a snake was coiled around him. The snake, in turn, disturbed from its own rest, struck out and bit him on the lip. Two other convicts, seeing what had occurred, came to the man's assistance 'and threw [the snake] on the ground, when it erected itself and flew at one of them; but they soon killed it'. But, by then the poison had begun to do its work, and by the following morning the man was dead. Collins is quick to tell us, however, that he died 'not ... from any effect of the bite of the snake, but from a general debility'.[15] He, like White, Tench and other commentators resisted the notion that Australian snakes were dangerous. In February 1795, Collins reinforced this by mentioning a non-fatal snakebite on a woman's leg,[16] but when a man from the Hawkesbury River area died after being bitten in December of the same year, there was no longer any question as to the cause of death.[17]

By the time of the first Australian newspaper report of a snakebite casualty, which appeared in the *Sydney Gazette* in 1804, the lethality of the colony's snakes had become well known. The report tells the story of a young boy at Hawkesbury who 'when sitting near a large tree [and] ... searching after the bandycoot, unhappily stretched one of his arms within the hollow, and suddenly withdrawing it much terrified' was bitten by a black snake. The report describes the child's immediate awareness of the likely outcome, stating how 'the poor little fellow, conscious of his danger, with an air of despondency remarked that he should soon die'. His death within twelve hours of the bite was not, by this stage, attributed to any constitutional feebleness or debility, but rather seen to be caused solely by the snake's venom. The report concludes: 'the body being examined, a wound appeared upon the left arm, thro' which the noxious viper had poured the contaminating fluid'.[18] The snake, the article implies, actively brought about the boy's death, by administering the fatal dose, while the boy had been instantly cognizant – and 'much terrified' – of the fatal nature of the bite. In the decade since the death of the sleeping convict, then, knowledge and acceptance of the toxicity of native snake venom had spread far into the settler community, so far that even young rural children by the banks of a river knew what grave consequences it brought.

So, unlike sharks, the risks posed by Australia's snakes had to be learned and absorbed; their threat to people who wandered beyond the encampments was not generally feared initially – meaning that tales of their presence could not be so readily used to police the 'wilderness' surrounding the settlements. Moreover, the surrounding bushland already had other 'dangers' that frightened the colonists and were invoked in warnings of the peril of overland escape – the Aboriginal population.[19] Well-rehearsed imperial fears about the 'natives' functioned as a similar deterrent to land escapes as sharks did in the sea. As John Hunter accounted in his 1793 published journal, stories abounded of convicts who fell victim to the 'savage outrages' of Aboriginal people. In one such breakout story, two convicts headed into the bush, the one escaped, but was wounded, the other has never been heard of since; but as some part of his cloaths were found which were bloody, and had been pierced by a spear, it was concluded he had been killed. A short time after this accident, a report prevailed, that part of the bones of a man had been found near a fire by which a party of the natives had been regaling themselves; this report gave rise to a conjecture, that as this man had been killed near this place, the people who had committed the murder had certainly ate him.[20]

Just as with sharks, tall tales of 'native savagery' played upon fears of the hidden predator, defiling the body and leaving nothing behind to be remembered or mourned, turning dreams of escape into nightmares. Even with such frightening tales circulating in the settlements, during the first few decades inland escapes were favoured by desperate runaways hoping to find outlaw camaraderie in their own band of 'merry men', or an overland route to China or Batavia (present-day Jakarta). Few were successful though and later breakouts favoured the sea.[21]

However, even if a snakebite was not the chief concern among convicts and colonists who ventured into the bush, it remained a grave possibility. Certainly a number of convicts who fled away from the coast died from envenomation and, like the stories of attacks by sharks or 'natives' they played a role in contributing to the useful idea of the 'wilderness' as hostile territory. A deadly snakebite was a fitting addition, for example, to the dastardly story of cannibalising Alexander Pearce who, in 1822, fled the notoriously brutal penal settlement at Macquarie Harbour in Van Diemen's Land with a group of seven other convicts. After a number of weeks, and two members of the group already murdered and consumed, Matthew Travers 'had his foot stung by some venomous reptile'.[22] The snakebite changed the dynamics in the tense relationships between the remaining men and, in the end, Travers died from a 'blow to the head' rather than the venom. But its place in the events rang powerfully in a narrative of the dangerous and villainous landscape beyond the camps, and was not, like his friends' cannibalism, a unique occurrence.

In proportion to the population, fatal encounters with dangerous animals were relatively more frequent than they are today, and snakes and sharks would have been seen in and around the settlements regularly. In addition, horror stories of escaping convicts, whose exploits were brought to a gruesome end, added to established fears. Together these quickly instituted the perception of an ominous and imminent danger posed by these animals within the first few decades of European arrival. Yet, as useful as these creatures were for the policing of boundaries around the settlements, invoked as fates worse than convictism, they did not passively and compliantly conform to the spatial framing imagined by the settlers. What began to concern the settlers as much as the perceived ubiquity of these dangerous creatures in the 'wilderness' beyond the settlements, and their apocryphal penchant for convicts, was the increasingly evident proximity of the 'danger' they posed within the camps and growing towns. These creatures increasingly displayed an ability to enter into the carefully cultivated and domesticated spaces of free and convict settlers alike. Much of the alarm expressed in the many reports of encounters with these animals during the first half-century and continuing into the later nineteenth century, was that they brazenly came into and threatened the controlled environment of house and home, workplace, garden, bathing areas, and so on. In 1795, seven years after his arrival in New South Wales, David Collins noted in his journal how a convict was bitten on the foot by a snake when he entered his hut, a space purposefully disconnected from the 'wilderness'. Collins noted that 'while we lived in the wood' – meaning before Sydney had been cleared – it was naturally 'expected to have been troubled with [snakes]' but after 'opening the country about us' he complained that snakes were still present in town, describing them as unwelcome intruders, exploiting the labours of Europeans, being 'often met in the different paths about the settlements, basking at mid-day in the sunshine'.[23] Snakes, Collins suggested, might reasonably be expected to reside within 'the woods', but they had no place in the cleared and reformed landscape of a burgeoning colony.

Thus, we might see Europeans in the penal colonies, as well as later free settlements, situated within a landscape of creatures that helped to set the spatial boundaries of settlement through the creation of risk barriers, but also trespassed on the carefully cultivated human environments of homes and gardens and streets. Not only were dangerous creatures like snakes and sharks 'out there' in the wilderness and deep ocean, defining the boundaries of both penal servitude and colony, but they were also 'right here', behind the woodpile, in the bathing places, where little fingers reached for lost toys, and where bare feet padded during the night toward the latrine. These animals intruded upon the 'civilised' domains and cultivated environs of European settlements, defying their attempts to tame and dominate the landscape. They were not cleared nor controlled but were recalcitrant agents that persisted in their claim to space.

By thinking about animals as not only passively affected by the course of colonisation, or as automatons onto which human responses and subjectivity could be imposed, we can begin to expand our understanding of how people moved around, created imperial spaces, and interacted with dangerous native animals as they carved homesteads, towns, cities, and prisons out of the wilderness.

As numerous scholars have shown, responses to and interactions with threatening species like these were not, of course, confined to Australia. Europeans in the penal and free settlements of Australia, and those who sought to explore its hinterlands and coast, did not confront their new surroundings, or the people and animals that inhabited them, as unique experiences or without reference to other colonial encounters. However, most of the scholarly literature that explores these interactions has focused on large carnivorous animals, such as lions, tigers, wolves, and to a lesser extent, crocodiles.

The latter, as well as sharks and dingoes, were the only large carnivores to inhabit Australia and its waters, and they, as alpha predators, raise interesting questions about mankind's claim to supremacy in the natural hierarchy.[24]

Perhaps of equal interest, although much less studied, are the less frequently discussed small, scuttling, and slithering creatures that hid in the woodpiles and under the floors of European homes and workplaces. These animals, like the jellyfish, cone shells, and blue ringed octopus whose deadly venom also made coastal activities potentially dangerous, were not included in imperial displays of masculinity and dominance, embodied in hunting expeditions and the collection of live and dead exotic trophies. As Harriet Ritvo explains in *The Animal Estate* (1987), one reason larger mammals were more central to symbolic displays of imperial enterprise was simply that they were easier to observe and interact with.[25] Their physiological structures were more familiar, and it was easy to place humans above them in the natural hierarchy. Reptiles, insects, and sea creatures were considered the lowest of animals, prehistoric and alien to such an extent that they appear marginal in the type of discourse and displays of empire that created zoological gardens and the exotic game parks described by John Mackenzie in *The Empire of Nature* (1988), for example. They came under the gaze of interested naturalists but unlike large carnivorous mammals were not celebrated in representations of colonial exploits, nor did they become symbols of European dominance over the environment, even when creatures like snakes and the 'Portuguese man o' war' were well known to pose significant danger to the humans who encountered them. And yet, they were more numerous, ubiquitous, and often in closer proximity to colonists as they built the homes, administrative buildings, and churches that helped consolidate claims to land throughout the world.

As an article in the *Sydney Gazette* from 1824 illustrates, animals like snakes were not rare or distant, requiring colonists to set out on expeditions to hunt or observe them, rather they trespassed on the private spaces and property of settlers in such a way as to undermine basic conceptions of 'civilisation' that the Europeans tried to establish through the clearing of land, planting of gardens, and building of homes. 'In Europe', the article reads, '[snakes] are not sufficiently numerous to be truly terrible. The philosopher can meditate in the fields without danger, and the lover seeks the grove without fearing any wounds but those of metaphor; not so in Australia, for even our firesides, fireplaces, and bed-chambers are occasionally visited by these poisonous intruders'.[26]

These interactions, in the most intimate environs of settler homes, work, leisure, and bathing places, were not simply alarming curiosities but events of frightening regularity. Innumerable cases appear in the archives in which a bite, or attack, is described within a context in which the offending creature should seemingly not belong – as if part of the horror of the encounter is in the incongruity of the place or activity with the dangerous animal. This seeming dissonance could be manifest in a number of different ways. The story of twelve-year-old Alfred Australia Howe, whose untimely death occurred along the Macleay River in northern New South Wales in 1837, is a good example. Various newspaper reports described how

> this unfortunate youth whilst washing his feet in shallow water, on the banks of the stream, in charge of a man servant, was suddenly seized by a large shark, near fifty miles from the harbour, and dragged into the current. The man rushed in and grasping the boy at the hazard of his own life, pulled him out of the monster's month and swam to land, just as the fish pursued them furiously to the shore.[27]

Unfortunately, however, the wound sustained during the ordeal was too great and while trying to get Alfred to proper medical care 'death terminated his sufferings by locked jaw, in a litter on the road'.[28] What was so troubling about the attack, aside from the sad irony that the boy had once before been 'saved from a watery grave' in a boating accident that had killed his father, was that the shark 'shouldn't' have been there. This was no swim in dangerous waters or capsized boat. It was neither the place, nor the activity that 'should' have put the boy in danger. Rather, he had been engaged in the simple domestic activity of washing, and he was at the shallow banks of a river 'fifty miles' from the sea. Not only was the attack unexpected, it had violated the space and activity of the boy's toilet, a decorous and 'human' pursuit. It was no place for an animal, let alone one that sought to overturn the natural hierarchy.

Similarly, in 1858 in the new suburb of Hawthorn on the edges of expanding gold-rush Melbourne, a woman called Margaret Maloney was bitten by what was identified in the newspapers as a 'whip snake'.[29] At the inquest, her husband, Michael, explained

> Margaret Molony was my wife – she was 29 years of age and we have five children. About eight o'clock ...[she] walked out of the back door about two yards from the house. I was talking to her when all of a sudden she jumped on one side and called out "Michael something bit me" – I saw a snake on the spot [she] left.[30]

The papers expressed the 'melancholy' nature of the death and emphasised the location of the assault 'not more than two yards' from the back door. Margaret, in conversation with her husband at the end of her day, had barely stepped beyond the safety of the threshold when the snake struck. She had not ventured out beyond the gate, or put her hand into the secret places where reptiles hide. The snake was in her space – 'on the spot she left' – and had not willingly surrendered it.[31]

While the statistical probability of these kinds of encounters was, and remains, low particularly when compared to other day-to-day dangers that increased with the growing population, such as accidents related to transportation and manual labour, their psychological impact was significant and became infused into the habits and cultural representations of the settler colony.

This is perhaps best and most famously characterised in Henry Lawson's 1892 short story 'The Drover's Wife' – now a canonical Australian text – in which a woman and her children who are 'nineteen miles to the nearest sign of civilization' realise that there is a snake beneath the floorboards of their house.[32] The story focuses on the isolation of the woman in her hut, raising and protecting her children in the harsh and lonely environment of the bush. The absoluteness of her situation, where the family is forced to spend the night in the kitchen, the mother sleeping on the dirt floor, is underscored by the hidden but menacing presence of the dangerous wildlife in the domestic home.

> She will not take [the children] into the house, for she knows the snake is there, and may at any moment come up through a crack in the rough slab floor; so she carries several armfuls of firewood into the kitchen, and then takes the children there. The kitchen has no floor – or, rather, an earthen one – called a "ground floor" in this part of the bush ... and then, before it gets dark, she goes into the house, and snatches up some pillows and bedclothes – expecting to see or lay her hand on the snake any minute. She makes a bed on the kitchen table for the children, and sits down beside it to watch all night.[33]

While appearing, on the face of it, to be vulnerable to the threat of the snake, the woman and her children, who Lawson tells us 'are Australian', bravely survive the encounter with the help of the loyal farm dog – the domesticated animal that 'rightly' shares a place alongside the family within the home. With fascinating irony, Lawson names the family dog 'Alligator', and it is he who alerts the woman to the approach of the snake as it ventures into the kitchen near dawn. The woman raises a stick to defend her children, and she and the dog work together in defeating the snake.

> [Alligator] has the snake now and tugs it eighteen inches. Thud, thud comes the woman's club on the ground. Alligator pulls again. Thud, thud. Alligator pulls some more. He has the snake out now – a black brute, five feet long. The head rises to dart about, but the dog has the enemy close to the neck. He is a big heavy dog, but quick as a terrier. He shakes the snake as though he felt the original curse in common with mankind... Thud, thud – its head is crushed, and Alligator's nose skinned... She lays her hand on the dog's head, and all the fierce, angry light dies out of his angry eyes.[34]

Brave encounters with Australia's dangerous animals began to appear with ever greater frequency in literature produced for both colony and metropole, in stories and novels written for adults and children from the middle of the nineteenth century. Beginning as narratives produced for English audiences and, particularly in the later decades of the nineteenth century, becoming more geared toward Australian readers, literary sources found a strong narrative focus and means of character development in encounters with dangerous animals. From the 1870s and 1880s, a new generation of literary and artistic production was exemplified by the founding of *The Bulletin*, a publication geared toward the creation of a new national self-image with the motto, 'Australia for the Australians'. In it, writers drew heavily on narratives of dangerous animal encounters in the fashioning of – primarily masculine – bushman heroes and anti-heroes. These characters demonstrated their self-reliance, energy, and toughness through an ability to survive life in the bush, which was often typified in their interactions with dangerous animals, even as Australians became more and more drawn to living in the growing cities.

Although anxious about the dangerous animals that were encountered in a variety of settings in the colonies, letter writers and diarists, even from the early decades of settlement, boasted of and often exaggerated the potency of the wildlife. Elizabeth Hawkins in writing to her sister about her journey through the Blue Mountains in 1822 grumbled that 'the greatest drawbacks of this country are the snakes which are so extremely venomous that no person who has been bitten has been known to live many minutes'.[35] Similarly, Felton Mathews, in his 1834 diary, alleged that the bite of a venomous snake would produce 'death in less than a quarter of an hour'. These snakes, he explained, were 'difficult to distinguish...from the fallen sticks among which they often lie'.[36] So, not only were they deadly but they were hidden from those without know-how or experience. Increasingly, the relationship between settlers and dangerous wildlife began to be presented not only as one of fear and a certain level of outrage, but also as one that demonstrated the bravery, hardiness, and toughness self-ascribed as defined characteristics of the Australian colonist. The animals are so dangerous – the narrative became – and can kill in minutes, and yet here we are living alongside them. Encounters with snakes and sharks became more casualised in the way that settlers wrote about them and interacted with them. John Sweatman, in his journal from the *HMS Bramble* that surveyed the coast of Australia in the 1840s, remarked on the carefree attitude of young Sydney-siders to the dangers of Sydney Harbour:

> Though there are plenty of sharks in some part of the harbour & I know of one being killed 17 feet in length...nobody seems to be afraid of them though for I constantly saw boys swim right across Wooloomooloo Bay to Garden Isd.[37]

The very creatures that had been employed as terrifying tools of custodial containment only decades earlier, now appeared absent from the enjoyments of a new generation that was distancing itself from its convict forbears and embracing their environment.[38]

So, the ways that European settlers and the generations that followed perceived and responded to these animals was not static or uniform. They ranged from exaggerated claims of their ferocity, bounties, eradication campaigns,[39] and day-to-day battles among settlers in trying to keep offending animals from crossing the borders of human-defined space, to their celebration and incorporation into a developing 'Australian' culture that cultivated the image of a uniquely robust and daring type of settler. Each manifestation reflected and determined the ways that people thought about and shaped their place within the environment.

But what of the animals to which the settlers were responding? Were they merely mirrors upon which colonial needs, anxieties, and identities were reflected, or did they play a part in the histories in which we find them? We can begin to examine the historical role of these creatures by interrogating not only the responses of Europeans to dangerous animals in this imperial setting but how these creatures appeared to respond in turn as their domain was changed and populated by more and more humans. The introduction of domesticated and invasive species of both animal and plant life, and the imposition of pastoral and agricultural technologies on the landscape, is increasingly appreciated as an important element of imperial history. Australia's dangerous creatures, however, challenge us to analyse the relationship of the invaders and settlers to the environment in reverse, by considering how they disrupted and contravened colonial endeavours to tame and clear the landscape. With the exception of the large carnivores, these animals continued to inhabit the new environments that were being created regardless of efforts to 'civilise' and cultivate them.

Coroners' inquests are filled with cases of fatal encounters that occurred within the home or workplace: where children crawled beneath furniture, where toes sought the dark corners of unattended shoes, or where youths joyfully splashed in shallow waters. In each of these encounters, in each historical vignette, human lives were ended or thrown into turmoil because of an encounter with an animal. And that animal life was affected. Frightened, startled, hungry, or aggressive, the animal in each case acted or reacted to human presence – the escaping convict stomping through the undergrowth; the fisherman struggling to right an upturned boat. For the animal, the encounter may be a moment of change too – it may have ended with a summary execution, painful wounds, or the satisfaction of a satiated hunger. Respectively, there were consequences for the animal as well as the human because of the encounter. Human and animal life was entangled in the moment and outcome of contact with one another.

During a mortal struggle with a predator such as a shark, both participants are drawn into each other's worlds, and each other's bodies, in a profoundly intimate manner.[40] I would also argue that the intimacy and particular horror of being taken into the body of an animal through predation corresponds strongly with the intimacy and horror of having something equally deadly and animal put within the human body. The injection of venom, a noxious animal fluid, travelling into the blood and tissue of a person, similarly blurs the corporeal boundaries of human and animal, such that the significant differences in human fatalities by predators and venomous animals are often ignored. Instead, in the popular imagination at least, deaths and injuries by 'deadly' wild animals of both categories are frequently brought together. In both, human (and animal) lives and bodies are most often profoundly changed and connected.

The predator or venomous animal is an active participant in the moment of contact and confrontation. 'Being eaten', historian Brett Walker tells us, 'is to confront our shared animal nature with other organisms, to surrender being human, a crafter of culture and artifice, and to regress into animal "barbarism"'.[41] The 'exceptionalism' of humans and their history is fundamentally undermined when a person is digested by another animal, leaving the cultural construction of human separation from nature gruesomely exposed, revealing the fragility of a human-centric reading of history.

Where then does that place animals within the narrative and analysis of history? Jason Hribal and Hilda Kean, for instance, have challenged us to go beyond merely inserting animals into existing historical frameworks and suggest rather that we should ask questions about their agency and role as historical 'actors'.[42] There is no doubt that in terms of their ability to 'make things happen', venomous animals, like snakes, spiders, and even jellyfish, contribute to and act within human stories. The power of their teeth or venom, that can infiltrate or tear apart the body of a person within minutes, rapidly alters biographical trajectories through death, scarring, or disability. That they intimately engage with humans, one to one, in morbid and mortal contact is clear but, as we have seen, they also engage with and act as protagonists in broader historical narrative.

From the beginning of European settlement in Australia, dangerous animals were called upon to participate in the policing of carceral boundaries. At the same time, they were feared and despised as they disrupted settler colonialist aims to replicate European society in lands appropriated from displaced First Peoples. They intruded upon the colonisers' fashioning of the landscape and entered their homes and private spaces, uninvited and unwanted. For this reason, they became instrumental in the construction of a colonial folklore that created a particularly hardy breed of settler who was able to withstand not only the dangers of the wilderness but also the perils that lived alongside them in their houses and cities. They were not merely beasts that were passively acted upon by the colonisers, but actors with competing claims for space, refusing to acquiesce to changes made to the landscape and sometimes denying humans their place at the top of the natural hierarchy.

Notes

1 N. Pateshall, *A short account of a voyage round the globe, performed in fifteen months*, State Library of Victoria (hereafter SLV), MS13479.
2 Robert Hughes, *The Fatal Shore: A History of the Transportation of Convicts to Australia, 1787–1868* (London: Harvill, 1996), 123.

3 James Bonwick, *Discovery and Settlement of Port Phillip: Being a History of the Country Now Called Victoria, Up to the Arrival of Mr. Superintendent Latrobe, in October, 1839* (Melbourne: George Robertson, 1857), 13.

4 'Garrison and General Order, Sullivan Bay, Port Philip [5]'. State Library of New South Wales (hereafter SLNSW) A341 [CY1151].

5 Watkin Tench, *A Complete Account of the Settlement at Port Jackson: In New South Wales, Including an Accurate Description of the Situation of the Colony; of the Natives; and of Its Natural Productions* (London, 1793), 107.

6 Jose I. Castro, 'On the Origins of the Spanish Word "Tiburon", and the English Word "Shark"', *Environmental Biology of Fishes*, 65 (2002), 249–53 at 250.

7 James Tuckey, *A Voyage to Establish a Colony at Port Philip in Bass's Strait On the South Coast of New South Wales, in His Majesty's Ship Calcutta, in the Years 1802–1804* (London, 1805), 33.

8 More information and sources about the dogs guarding Port Arthur can be found at Margaret Harman, 'Dog-Line at Eaglehawk Neck' in Alison Alexander (ed.), *The Companion to Tasmanian History* (Tasmania: University of Tasmania, 2006), www.utas.edu.au/library/companion_to_tasmanian_history/D/Dog-line%20at%20Eaglehawk%20Neck.htm, accessed 19 Sep. 2016.

9 Ian Brand, *Penal Peninsula: Port Arthur and its Outstations, 1827–1898* (Launceston: Regal Publications, 1989), 112.

10 Godfrey Charles Mundy, *Our Antipodes: or, Residence and Rambles in the Australasian Colonies. With a Glimpse of the Gold Fields* (London: R. Bentley, 1852), 505.

11 Stoney H. Butler, *A Residence in Tasmania: with a Descriptive Tour Through the Island, from Macquarie Harbour to Circular Head* (London: Smith, Elder & Co., 1856), 54.

12 Jane A. Fletcher, *Eaglehawk Neck: A Brief History of its Military Occupation Compiled from Various Notes* (Hobart: Mercury Press, 1946), 7. [Primary source confirmation of the use of offal to attract sharks is elusive, but Anne Hoyle, Resource Centre Officer for the Port Arthur Historic Site Management Authority, as well as Professor Hamish Maxwell-Stewart of the University of Tasmania, and independent scholar Simon Barnard, have all confirmed its likelihood in various email correspondence during Sep. 2014.]

13 John White, *Journal of a Voyage to New South Wales with Sixty-five Plates of Nondescript Animals, Birds, Lizards, Serpents, Curious Cones of Trees and Other Natural Productions* (London, 1790), 259.

14 Tench, *A Complete Account of the Settlement at Port Jackson*, 177.

15 David Collins, *An Account of the English Colony in New South Wales: With Remarks on the Dispositions, Customs, Manners, &C. of the Native Inhabitants of that Country* (London, 1798), 198.

16 Ibid., 406.

17 Ibid., 441.

18 "Sydney", *The Sydney Gazette and New South Wales Advertiser*, 21 Oct. 1804, 2.

19 Grace Karskens, *The Colony: A History of Early Sydney* (Sydney: Allen and Unwin, 2009), 362.

20 John Hunter, *An Historical Journal of the Transactions at Port Jackson and Norfolk Island* (London, 1793), 67.

21 'Manuscript on Convict Escapees, 1830'. National Library of Australia (hereafter NLA), MS 9438; see also Hughes, *The Fatal Shore*, 203–204.

22 Alexander Pearce, *Narrative of the Escape of Eight Convicts from Macquarie Harbour in Sept, 1822, and of the Their Murders and Cannibalism Committed During Their Wanderings*. SLNSW, DMLS 3.

23 Collins, *An Account of the English Colony in New South Wales*, 282.

24 See, for example, John M. Mackenzie, *The Empire of Nature: Hunting, Conservation, and British Imperialism* (Manchester: Manchester University Press, 1988).

25 Harriet Ritvo, *The Animal Estate: The English and Other Creatures in the Victorian Age,* (Cambridge, MA: Harvard University Press, 1987), 15.

26 'Police Office News', *The Sydney Gazette and New South Wales Advertiser,* 18 Nov. 1824, 3.

27 'Family Notices', *The Sydney Monitor,* 1 Feb. 1837, 3.

28 'Family Notices', *The Sydney Gazette and New South Wales Advertiser,* 31 Jan. 1837, 3.

29 'The Suburban Railway', *The Argus,* 5 Mar. 1858, 5 & 'Inquests', *The Age,* 9 Mar. 1858, 5. The snake was described as 'about two feet ten inches in length, and was of a greenish color' and although being identified in the newspapers as a 'Whip Snake' was probably more likely to have been an Eastern Brown Snake. The Whip Snakes that are found around Melbourne do not produce venom that is fatal to adult humans. They also have dark coloured heads that would likely have been an identifying feature. The Eastern Brown is extremely venomous and capable of killing an adult woman, like Margaret Moloney, within twelve hours. While it is brown, it could be seen to have a greenish hue.

30 Public Record Office of Victoria (hereafter PROV), VPRS 24/P/0, Unit 60, Item 1858/63.

31 After tending to her wound, Michael left Margaret for a few minutes to go back to the yard and kill the snake 'which it was trying to get under the house'. Ibid.

32 Henry Lawson, 'The Drover's Wife', *While the Billy Boils* (Sydney: Angus and Robertson, 1915 [1896]), 169–80 at 169.

33 Ibid., 170–71.

34 Ibid., 179–80.

35 Letter, Elizabeth Hawkins to 'Ann', dated Bathurst, 7 May 1822. NLA, MS 535.

36 Entry of 14 Jan. 1834 in Diaries of Felton and Sarah Mathews, 1829–1834. NLA, MS 15.

37 'Voyage to N.E. Australia – M.S. Journal – Sweatman, 1848'. SLNSW, Microfilm A 1725.

38 John Bigge, 'Report of the Commissioner of Inquiry on the State of Agriculture and Trade in the Colony of New South Wales', Parl. Papers (Commons), 1823 (136).

39 Lady Jane Franklin, wife of John Franklin who became lieutenant-governor of Van Diemen's Land in the 1830, established a bounty scheme for ridding the island of snakes. She offered one shilling for every snake's head brought to her and reportedly spent over £600 in one season killing thousands of snakes and throwing penal labour and discipline into turmoil.

40 Brett Walker, 'Animals and the Intimacy of History', *History and Theory,* 52/4 (Dec. 2013), 45–67.

41 Ibid., 54.

42 Hilda Kean, 'Challenges for Historians Writing Animal-Human History: What Is Really Enough?' *Anthrozoös,* 25 (2012), S57–72; and Jason Hribal, 'Animals, Agency, and Class: Writing the History of Animals from Below', *Human Ecology Review,* 14 (2007), 101–12.

Bibliography

Primary Sources

Archival Material

National Library of Australia, Canberra.
    MS 15. Diaries of Felton and Sarah Mathews, 1829–1834.
    MS 535. Typescript Copy of Letter of Elizabeth Hawkins, 7 May 1822.
    MS 9438. Manuscript on Convict Escapees, 1830.

Public Records Office of Victoria, Melbourne.
  VPRS 24/P/0, Unit 60, Item 1858/63. Inquest Deposition File.

State Library of New South Wales, Sydney.
  A341 [CY1151]. Garrison and General Order, Sullivan Bay, Port Philip [5].
  Microfilm A 1725. Voyage to N.E. Australia – M.S. Journal – Sweatman, 1848.
  DMLS 3. Pearce, Alexander, 'Narrative of the Escape of Eight Convicts from Macquarie Harbour in Sept., 1822, and of the Their Murders and Cannibalism Committed During Their Wanderings'.

State Library of Victoria, Melbourne.
  MS13479. Pateshall, N. 'A short account of a voyage round the globe, performed in fifteen months'.

Newspapers and Periodicals

The Age
The Argus
The Sydney Gazette and New South Wales Advertiser
The Sydney Monitor

Other Published Material

Bigge, John, 'Report of the Commissioner of Inquiry on the State of Agriculture and Trade in the Colony of New South Wales', Parl. Papers (Commons), 1823 (136).
Collins, David, *An Account of the English Colony in New South Wales: With Remarks on the Dispositions, Customs, Manners, &C. of the Native Inhabitants of that Country* (London, 1798).
Hunter, John, *An Historical Journal of the Transactions at Port Jackson and Norfolk Island* (London, 1793).
Mundy, Godfrey Charles, *Our Antipodes: Or, Residence and Rambles in the Australasian Colonies. With a Glimpse of the Gold Fields* (London: R. Bentley, 1852).
Stoney, H. Butler, *A Residence in Tasmania: With a Descriptive Tour Through the Island, from Macquarie Harbour to Circular Head* (London: Smith, Elder & Co., 1856).
Tench, Watkin, *A Complete Account of the Settlement at Port Jackson: In New South Wales, Including an Accurate Description of the Situation of the Colony; of the Natives; and of Its Natural Productions* (London, 1793).
Tuckey, James, *A Voyage to Establish a Colony at Port Philip in Bass's Strait on the South Coast of New South Wales, in His Majesty's Ship Calcutta, in the Years 1802–1804* (London, 1805).
White, John, *Journal of a Voyage to New South Wales with Sixty-five Plates of Non-descript Animals, Birds, Lizards, Serpents, Curious Cones of Trees and Other Natural Productions* (London, 1790).

Secondary Material

Bonwick, James, *Discovery and Settlement of Port Phillip: Being a History of the Country Now Called Victoria, Up to the Arrival of Mr. Superintendent Latrobe, in October, 1839* (Melbourne: George Robertson, 1857).

Brand, Ian, *Penal Peninsula: Port Arthur and Its Outstations, 1827–1898* (Launceston: Regal Publications, 1989).

Castro, Jose I., 'On the Origins of the Spanish Word "Tiburon", and the English Word "Shark"', *Environmental Biology of Fishes*, 65 (2002), 249–53.

Fletcher, Jane A., *Eaglehawk Neck: A Brief History of Its Military Occupation Compiled from Various Notes* (Hobart: Mercury Press, 1946).

Harman, Margaret, 'Dog-Line at Eaglehawk Neck' in Alison Alexander (ed.), *The Companion to Tasmanian History* (Tasmania: University of Tasmania, 2006), www.utas.edu.au/library/companion_to_tasmanian_history/D/Dog-line%20at%20Eaglehawk%20Neck.htm, accessed 19 Sep. 2016.

Hribal, Jason, 'Animals, Agency, and Class: Writing the History of Animals from Below', *Human Ecology Review*, 14 (2007), 101–12.

Hughes, Robert, *The Fatal Shore: A History of the Transportation of Convicts to Australia, 1787–1868* (London: Harvill, 1996).

Karskens, Grace, *The Colony: A History of Early Sydney* (Sydney: Allen and Unwin, 2009).

Kean, Hilda, 'Challenges for Historians Writing Animal-Human History: What Is Really Enough?' *Anthrozöos*, 25 (2012), S57–72.

Lawson, Henry, 'The Drover's Wife', *While the Billy Boils* (Sydney: Angus and Robertson, 1915 [1896]), 169–80.

Mackenzie, John M., *The Empire of Nature: Hunting, Conservation, and British Imperialism,* (Manchester: Manchester University Press, 1988).

Ritvo, Harriet, *The Animal Estate: The English and Other Creatures in the Victorian Age,* (Cambridge, MA: Harvard University Press, 1987).

Walker, Brett, 'Animals and the Intimacy of History', *History and Theory*, 52/4 (Dec. 2013), 45–67.

Afterword

Harriet Ritvo

In recent decades, scholarship focused on other animals and especially on the relation between people and other animals, has proliferated throughout the humanities and social sciences. This flowering of academic interest has been acknowledged and institutionalized in conventional academic ways, which is to say, at first in conferences, journals, book series, and ultimately in degree programs and faculty positions. Such programs and positions tend to be described as interdisciplinary, a quality that seems implicit in their most frequent labels: 'animal studies' or 'human-animal studies'. Often, however, such formal recognition, or the increase in interest and prestige that it signals, has had a somewhat paradoxical effect. As animal-related topics have become respectable (which has not always been the case), they have been mainstreamed in various disciplines; in consequence, such research has increasingly reflected disciplinary conventions. In addition, as a critical mass of specialists has been reached in field after field – enough to sustain a lively conversation – they have predictably inclined to talk among themselves. This inclination can be seen even within animal studies itself. Although diverse scholars find shelter beneath its broad umbrella (and attend its conferences and contribute to its journals), the center of disciplinary gravity tends toward literary studies and philosophy. One result – doubtless collateral damage rather than intentional exclusion – has been the marginalization of historical approaches.

Of course, this development should not be surprising. Despite occasional claims to the contrary, most scholars work within one discipline or another. The acknowledgement throughout the humanities and (at least many of the) social sciences that other animals merit serious scholarly consideration arguably marks a more significant achievement than does the emergence of animal studies as a separate area of study. (That is to say, it makes more difference in the experience of academics – not necessarily in the experience of their non-human subjects.) But, as *Interspecies Interactions* demonstrates, disciplinary boundaries can be productively permeable. Though the focus of the volume is clearly historical, and historians predominate among the contributors, allied disciplines including literary studies, art history, and archaeology are illuminatingly represented. By implication, the readership should be at least equally broad.

Taken as a group, the essays collected here show the vitality of historical research on animal topics, the extent to which historians have integrated insights and methods from other fields, and the way that historical perspective can enrich work in fields closer to the core of animal studies. They show how conventional historical sources can yield fresh insights when they are interrogated with other animals in mind, whether a horse whose life was particularly well documented or the creatures who suffered on the way to the table (or even at the table). They complicate understandings of such well-studied topics as the intellectual history of the Enlightenment and the origins of national identity. They retrieve the intertwined experiences of human and non-human animals, in settings from Italian courts to Ottoman carnivals to English sporting arenas. And they provide very usefully grounded explorations of the agency of animals, whether the varied objects of (real or imagined) Scottish lust or the equine subjects of German manuals or the jellyfish, snakes, and sharks who helped shape colonial impressions of Australia.

The treatment of agency, which has received a great deal of attention within the field of animal studies, exemplifies the special contribution that historians have to make. The term "agency" has eluded precise definition or, to put it another way, it is subject to many alternative definitions. Perhaps for that reason, it often figures as an abstraction. Anchoring this vexed general issue in particular times, places, and bodies can make the stakes much clearer. And the same could be said of the term 'animal,' which in its most expansive sense encompasses organisms so diverse as to defy generalization, or of 'the animal,' which therefore works to reinforce the human-animal binary that much animal studies scholarship seeks to abrogate. Humans are ineluctably omnipresent in animal-related scholarship; history can ensure that real animals are there too.

Index

Note: Regnal and vital dates are listed only in certain cases to avoid ambiguity.

Made in the USA
Middletown, DE
05 January 2020